THE GREATEST SCIENTIFIC GAMBLE

THE GREATEST SCIENTIFIC GAMBLE

A Story of Impossible Odds, Rival Scientists, and the Atomic Bomb

Michael Joseloff

East Lansing • *Michigan State University Press*

Michigan State University Press
East Lansing, Michigan 48823-5245

Library of Congress Cataloging-in-Publication Data is available
ISBN 978-1-61186-560-8 (cloth)
ISBN 978-1-61186-559-2 (paper)
ISBN 978-1-60917-801-7 (PDF)
ISBN 978-1-62895-564-4 (ePub)

Book design by Anastasia Wraight
Cover design by David Drummond
Cover art is First nuclear reactor at the University of Chicago,
Chicago Pile-1 CP-1, Shutterstock.

Visit Michigan State University Press at *www.msupress.org*

To Joanne, Matthew, and Karen

CONTENTS

PREFACE

Several years ago, while browsing the internet, I came across a photo taken at a University of Michigan physics conference in September 1939. It showed Werner Heisenberg, the future architect of Germany's atomic research program, standing with Enrico Fermi, a future Manhattan Project scientist, and Sam Goudsmit, who would go on to head an army intelligence unit tasked with gathering information on Germany's atom bomb program and, at the end of the war, capturing Heisenberg. I was surprised to see them together one month before the launch of World War II. I subsequently learned that the three were longtime friends and that Heisenberg had paid a visit to another friend that summer, Robert Oppenheimer.

The invention of the atom bomb was one of the greatest scientific and engineering feats of the twentieth century and, according to President Truman, "the greatest scientific gamble in history." I became interested in that "gamble" years ago while researching a TV news report on Oppenheimer. That photo rekindled my interest. Today everyone knows how that "gamble" turned out. Those men were principal players and that photo suggested a compelling way to tell their story. *The Greatest Scientific Gamble* weaves the lives of those friends-turned-rivals with that of General Leslie Groves, director of the Manhattan Project, into a high-stakes narrative about Allied scientists and engineers confronting unprecedented, seemingly insurmountable obstacles in a race that could determine the outcome of World War II; a race to beat Heisenberg to an atom bomb.

CHAPTER ONE

A SMALL GLOBAL COMMUNITY

CBS RADIO NEWS, MARCH 13, 1938

Tonight the world trembles. Right at this moment Austria is no longer a nation, but is now officially a part of the German empire . . . The Nazis are driving with all their might to bring Austria under complete Nazi domination . . . Jews, Catholic leaders and former Austrian officials are being jailed. Hitler . . . is preparing . . . a roundabout triumphal tour of the land of his birth . . . In Berlin, Field Marshal Herman Goering has served notice that Germany intends to go after the Germans in Czechoslovakia already ringed on several sides by German troops.

By the summer of 1939, Adolf Hitler had all but shredded the Versailles Treaty, the peace agreement ending World War I. While Germany's erstwhile enemies sat idly by, he'd rebuilt his nation's military, sent German troops into the demilitarized Rhineland, annexed Austria, bullied the Czechs into surrendering the Sudetenland, occupied the rest of Czechoslovakia, and stripped German Jews of their civil rights, forcing many to flee the country.

After years of grabbing territory through political chicanery, Adolf Hitler was ready to go to war; his plan: to create a Third Reich, a racially pure Germanic Empire in Europe. As the date for his opening attack on Poland neared, scientists in Germany, Russia, France, Britain, and the United States were feverishly conducting experiments in nuclear fission, a newly discovered field in which the nucleus of a uranium atom is split and emits a powerful burst of atomic energy. By the end of 1939, more than a hundred articles on the subject had appeared in scientific journals.[1] But the German War Office was the first to fund research for a weapon using atomic energy, a bomb that would draw its power not from conventional manmade explosives like TNT but from a primordial energy locked

Samuel Goudsmit, Clarence Yoakum, Werner Heisenberg, and Enrico Fermi at the University of Michigan, summer 1939. COURTESY AIP EMILIO SEGRÈ VISUAL ARCHIVES, CRANE-RANDALL COLLECTION, GOUDSMIT COLLECTION.

deep inside the uranium atom. No one had ever tried to harness that energy. No one knew if it could be done. But one atomic bomb could theoretically destroy an entire city. Loaded into a long-range strategic bomber or onto a guided rocket, it could terrorize nations, pressuring them to surrender and spread Nazism across the globe.

With the world on the brink of war, Werner Heisenberg, a thirty-seven-year-old professor of theoretical physics at the University of Leipzig, a Nobel Prize winner, and future architect of Hitler's atom bomb program boarded an ocean liner bound for New York City. From there he would go on to Ann Arbor, Michigan, to attend an annual University of Michigan physics conference.

Heisenberg was one of the most brilliant physicists of his time. His work in Matrix Mechanics was considered as revolutionary as that of Albert Einstein's theory of relativity. A corporal in the reserve force that supported Germany's regular army and a patriot, Heisenberg sympathized with the Nazi goal of restoring Germany to a position of prominence in Europe.[2] But he decried what he called "the excesses" of Nazism, especially the anti-Semitic policies that drove Jewish scientists, some of the nation's best, to flee the country.[3]

At the time, there were only a couple hundred scientists in the world doing research in theoretical physics.[4] "We all felt that we belonged to a sort of lodge with a worldwide membership," commented Sam Goudsmit, a University of Michigan physics professor—a lodge where everyone knew each other or were at least familiar with their work.[5] With the rise of Nazism, Heisenberg found himself estranged from that small global community, and he looked forward to spending time with old friends in Michigan, some of whom he had known since the 1920s when they were postdoctoral fellows and he was a teaching assistant in Göttingen, Germany. Among them: a thirty-four-year-old physicist named J. Robert Oppenheimer.[6]

Göttingen was the world capital of theoretical physics. Oppenheimer, a Harvard graduate and a member of a wealthy New York City Jewish family, had come there to study in 1927. He'd read and admired Heisenberg's work and looked forward to meeting him. Heisenberg, as it turned out, was teaching elsewhere at the time, but when he came back to campus for a brief visit, Oppenheimer sought him out and introduced himself. Both men had brilliant and original minds. Drawn together by their shared interest, the future rivals developed a collegial relationship.

The University of Michigan seminars drew top theoretical physicists from all over the world. Heisenberg had attended several times in the past and always stayed at the home of his good friend Samuel Goudsmit, a Dutch American Jew who had emigrated to the United States in 1927. While a postgraduate student at the University of Leiden in the Netherlands, Goudsmit had published a paper on electron spin that led to a job offer from the University of Michigan, which was starting a theoretical physics department. At the time, Europeans considered the scientific research coming out of American universities to be uninteresting and boring, but Goudsmit and his wife were living in a single room in a rooming house without a bath or kitchen.[7] He needed a job that paid a decent salary, so he reluctantly accepted.[8] When their ship docked in New York Harbor, Robert Oppenheimer was there to greet them and help them get situated as they headed on to Michigan.[9]

In Michigan, Heisenberg reconnected with a second former Göttingen student, Enrico Fermi, a Nobel Prize–winning Italian physicist. Fermi had won a prestigious Italian government scholarship to pay his way at the elite German university. Even so, his seven months there had been difficult. The son of an

Italian railroad administration bureaucrat, Fermi was dogged by insecurity and fears of not measuring up. Despite such worries, he and Heisenberg became close friends.[10]

Heisenberg's trip to the 1939 conference would be his last. The Japanese attack on Pearl Harbor in December 1941 would thrust him and his three American friends into the front lines of World War II where they would square off against each other in a race to build an atomic bomb.

America's entry into the war came almost four decades after Albert Einstein revolutionized atomic physics with his theory of special relativity that energy and mass were one and the same. An atom's mass was simply energy in a different form. In the years since, his famous equation $E=mc^2$ had become part of popular culture. Publishers lured readers with sensational headlines highlighting the potential dangers of atomic energy: "Tinkering with Angry Atoms May Blow Up the Earth" and "Nature's Greatest Secret, the Might of the Atom, Stirs Scientists."[11]

In 1914, in his novel *The World Set Free*, H. G. Wells looked into his crystal ball and conjured the horror of 1959, the year the world's nations became engulfed in an atomic war. "From nearly 200 centres . . . roared the unquenchable crimson conflagrations of the atomic bombs," he wrote, "the flimsy fabric of the world's credit had vanished, industry was completely disorganised and every city, every thickly populated area was starving or trembled on the verge of starvation."[12]

When Wells wrote his apocalyptic novel, atomic energy was a mysterious, almost mythical force of nature. Hidden in the inner sanctum of the atom's nucleus and protected from inquisitive scientists by multiple layers of defense, it was virtually impossible to reach. Impossible, that is, until Enrico Fermi began chipping away at those defenses.

CHAPTER TWO

A MODERN ALCHEMIST

Scientific understanding of radiation and atomic structure had advanced dramatically over the past decades. Electrons with their negative electrical charge had been discovered in the late 1890s, the atomic nucleus in 1911, and the positively charged protons inside the nucleus in 1919, but it was the 1932 discovery of an additional particle inside the nucleus, neutrons with no electric charge, that paved the way for a still greater discovery.[1]

"Clear the way. Look out!" shouted Enrico Fermi as he exited his laboratory clutching an irradiated sample of aluminum and sprinted to a Geiger counter in a lab down the hall. It was winter 1934. Fermi, a physics professor at the University of Rome, was attempting some modern alchemy, bombarding samples of aluminum and other non-radioactive elements with neutrons, trying to turn them radioactive.[2]

The radiation would last just a short while, often less than a minute. To make certain that errant neutrons didn't contaminate the results, he placed the neutron source and the Geiger counter that measured the intensity of the induced radiation at opposite ends of the building. Thus, the thirty-three-year-old Fermi and his postgraduate students racing like crazed schoolboys down the hall became a familiar sight all during that winter, spring, and fall.[3] The sprinter was often the professor himself. Fermi prided himself on his fleetness of foot and his ability to out-run, out-hike, and out-swim his young associates.[4]

The inspiration for Fermi's experiment came from two French scientists, Frédéric and Irène Joliot-Curie, daughter of Marie Curie, who had bombarded atoms with positively charged particles. Atoms had a positively charged outer barrier that repelled positively charged particles, so their experiments could only induce radiation in lighter atoms—those with weaker barriers.[5] Fermi reasoned that neutrons which have no electric charge could penetrate even the strongest defenses.

To produce those neutrons, Fermi needed two elements: radium, which emitted radon gas when it decayed, and beryllium. At the time, a gram of radium cost 670,000 lire (about 34,000 dollars), far more than a university physics department could afford.[6] Fortunately, Rome's Bureau of Public Health was housed in the physics department building and kept a gram of radium locked in a basement safe. Happy to help, the Bureau's director allowed Fermi and associates to siphon off as much radon as they needed.

To prepare for their experiment, Fermi's students visited Mr. Troccoli, owner of Rome's largest chemical supply store.[7] There, they purchased samples of beryllium and some sixty other elements; among them aluminum, carbon, platinum, and gold. Back in the lab, they began the tedious process of irradiating each sample. One by one they wrapped them around a glass tube containing radon gas and beryllium powder and placed the whole in a lead-lined box.[8]

If all went as planned, the radon atoms would slam into the beryllium atoms, causing their nuclei to split and emit neutrons (more than 100,000 per second).[9] The freed neutrons would then penetrate the atomic nuclei in the test sample, rendering them unstable so that they would decay and emit radiation.[10] The final test targeted uranium, which was naturally radioactive, so no one was certain how it would react, or if it would react. To find out, Fermi and his colleagues performed a chemical analysis following the bombardment. Uranium atoms had ninety-two protons in their nuclei. At the time, uranium was the largest, heaviest known element on earth. So, Fermi was shocked when he discovered atoms with ninety-three protons.

The bombarded uranium atoms had apparently undergone a nuclear makeover. No longer uranium, they had gained a proton and been transformed into a heavier "transuranic" element. Not yet willing to accept the results, Fermi performed several tests to verify their accuracy. Only then, treading cautiously, did he and his students co-author a research paper reporting that the evidence *suggested* the possibility of a historic breakthrough, the creation of a new manmade element with ninety-three protons.[11] The director of Fermi's Institute at the University of Rome was not as cautious. He announced unequivocally that Enrico Fermi was the first scientist to create a manmade atom. The Italian press boasted it was a "fascist victory of the cultural arena."[12] Other news outlets took notice. The *New York Times* trumpeted Fermi's achievement with the headline "Italian Produces 93rd Element by Bombarding Uranium."[13]

Fermi's neutron bombardment experiments precipitated a new era of nuclear exploration. Some thought neutron bombardment held the key to unlocking atomic energy. But such speculation was quickly dismissed. The Nobel laureate Ernest Rutherford called the idea "moonshine." Albert Einstein concurred, saying that using neutron bombardment to harness atomic energy would be like "shooting in the dark at scarce birds."[14] Even so, it was clear that something strange and exciting was going on within the nuclei of bombarded atoms. And the surprises kept coming.

In October 1934, Fermi was heading home from a lecture tour in South America when his colleagues telegraphed him about a problem. They had bombarded aluminum in the lead-lined box, then bombarded it again to verify the results. The two experiments produced different numbers. The intensity of the artificial radioactivity seemed to vary from experiment to experiment. Determined to find out what was going on, they designed a new test. In one experiment, they placed an aluminum sample and neutron source in a corner of the box; in a second experiment, they placed them in the center. To their surprise, they got different results. Similarly, an experiment conducted on a wooden table turned out differently than the same experiment laid out on a table with a marble top.[15] The divergent findings meant that their months of work were now in question. Fermi was not pleased, and he let his young associates know it.

One day after Fermi's return, his associates set up an experiment to see what would happen if a piece of lead were inserted between the neutron source and a piece of silver. They were out of the lab, proctoring student exams, when Fermi wandered in and decided to go ahead with the experiment. On a whim, he removed the lead and substituted a piece of paraffin wax. "It was just like that," he would later comment, "with no advance warning, no conscious prior reasoning."[16]

After conducting his experiment, Fermi grabbed the silver sample and sprinted down the hall to the Geiger counter as he had done dozens of times before. But these results were unlike any he'd ever seen. The radioactivity level soared to an intensity more than a hundred times greater than in any previous experiment. When his young colleagues arrived, they repeated the experiment. And, according to Enrico's wife Laura, there was pandemonium. The halls of the physics building resounded with loud exclamations: "Fantastic! Incredible! Black magic!"[17]

Fermi theorized that the neutrons had collided with the hydrogen atoms in the paraffin. The ricocheting back and forth had slowed down the speeding neutrons so that they spent more time in the vicinity of the target atoms and had more time to induce a reaction.[18] The marble and wooden tabletops in the previous experiments and the walls of the lead-lined box had apparently had a similar effect.

To test his theory, Fermi led his excited young colleagues out behind the physics building to a private garden owned by the director of the physics department. Its centerpiece was a fishpond stocked with goldfish. Fermi lowered his neutron source and a cylinder containing silver into the water. To their delight, they discovered that inserting water (H_2O), with all its hydrogen atoms, between the neutron source and the piece of silver also produced a burst of artificial radioactivity.

CHAPTER THREE

NOBEL PRIZE

Early on the morning of November 10, 1938, the telephone rang in the Fermi's spacious Rome apartment. Laura Fermi rushed into the hall to answer it. "I wish to inform you," the caller said, "that this evening at six, Professor Fermi will be called on the telephone from Stockholm."[1]

Fermi had been tipped off a month earlier, while attending a conference in Copenhagen, to expect the call. Understandably excited, Laura asked her husband to take the day off. The couple went on a shopping spree. They bought expensive watches. She bought a beaver skin coat. Unlike diamond purchases that were recorded and might draw official scrutiny, time pieces and warm coats did not leave a paper trail. Equally important, they could be resold if for some reason the family needed cash.

The telephone rang as they walked in their apartment. It was a friend, calling to see if the Stockholm call had come through. At six o'clock, still waiting, they turned on the radio. The news was not good. Italy's dictator Benito Mussolini had just announced new restrictions on the rights of Italian Jews. Enrico was Catholic and their children were being raised Catholic, but Laura was Jewish. Jews made up only a tiny fraction of the population, and Italians, for the most part, were not anti-Semitic. Still, months earlier, Mussolini, following in the footsteps of Adolf Hitler, had launched a campaign against Italy's Jews.[2] Now, it seemed that every day brought more bad news. This time, the commentator announced laws excluding Jewish children from public schools, dismissing Jewish teachers, and restricting Jewish professionals from working for non-Jewish clients. To make matters worse, authorities announced that the passports of Italian Jews would soon be confiscated.

Laura and Enrico had lived a comfortable professorial life up until then: a nice apartment in Rome, summers at a cabin in the Alps, interesting travel with lecture tours in South America and the United States. Now, looking ahead, they saw only hardship and despair. They had planned to abandon their home and leave Italy early in the new year, but now it seemed that time was running out.

Shortly after six o'clock, the telephone bell rang again. It was the secretary of the Swedish Academy of Sciences in Stockholm. Enrico took the phone, and the woman on the other end read him the citation on his Nobel Prize: for his "identification of new radioactive elements produced by neutron bombardment and his discovery . . . of nuclear reactions effected by slow neutrons."[3]

When the news of Fermi's Nobel Prize got out, friends rushed to his apartment to join in the celebration. But Mussolini's latest pronouncement had left Laura shaken. She wrote later, "I did not know whether to be happy or sad, whether to heed the telephone or the radio."[4] Enrico, his usual confident self, assured her that everything would work out.

The timing of the Nobel Prize was a godsend, but the stamp on Laura's passport identifying her as a Jew was a problem. To reach Stockholm, the Fermis would have to take a train through Germany, and border guards would undoubtedly check their documents.[5]

Desperate times called for desperate measures. Enrico contacted their local parish priest and made plans to baptize Laura and their two children into the Catholic Church and give him and Laura a Catholic wedding.[6] The passport issue resolved, the Fermis were planning their escape when, in Germany, fanatical anti-Semitism erupted into widespread violence.

On the night of November 9, 1938, and into the following morning, Nazi mobs across Germany burned and vandalized Jewish synagogues, homes, and businesses. They killed one hundred Jews, arrested thirty thousand more, and sent them to concentration camps. The paroxysm of violence known as the "Night of Broken Glass, or "Kristallnacht" in German, was triggered when a seventeen-year-old Polish Jew protesting the expulsion of Polish Jews from Germany assassinated a German diplomat. German anti-Semitism had inspired localized mayhem before, but Kristallnacht was the first government orchestrated nationwide outbreak.

On December 6, 1938, Enrico, Laura, their young son, daughter, and nursemaid, and a few close friends waited at Rome's Termini station for the train that would take the Fermis to Stockholm and eventually to their new home in the United States. It was a highly emotional moment. Ginestra Amaldi, wife of one of Enrico's postgraduate students and a close friend of Laura, could not help herself.

Laura, Giulio, Nella, and Enrico Fermi. COURTESY AIP EMILIO SEGRÈ VISUAL ARCHIVES, WHEELER COLLECTION.

She felt abandoned and angry. "Enrico's departure," she said, "is a betrayal of the young people who have come to study with him and who have trusted him for guidance and help."[7] Her husband defended his mentor and friend. Laura, already conflicted because she was leaving her sisters and father behind, was deeply hurt.

Four days later, Enrico sat on the stage at the Royal Opera House waiting to shake hands with the King of Sweden and accept his medal along with a large monetary award. The handshake did not go over well back home. Mussolini and his supporters expected a raised arm, the Fascist salute. They interpreted the handshake as an act of defiance. But the Fermis were beyond the reach of Italian authorities.[8] They used the prize money to book passage on an ocean liner bound for New York City. Anticipating the family's need to escape, Enrico had accepted a job offer at Columbia University.

The fermi family had been in New York for just two weeks, looking for an apartment and getting settled, when the liner *Drottningholm* sailed into New York harbor. On an upper deck peering down at the welcoming crowd was the Danish Nobel laureate Niels Bohr. A luminary in the theoretical physics community and a mentor to Heisenberg, he brought with him the incredible news that German scientists had proven Einstein's theory correct. Otto Hahn and Fritz Strassmann had split the nucleus of a uranium atom, unlocking a mighty storehouse of atomic energy. And they had done it using the experiment Fermi had pioneered: bombarding uranium with slow-moving neutrons.

The news left Fermi uneasy. His thoughts raced back to his own neutron bombardment experiments at the University of Rome five years earlier. He remembered the anxiety he felt when the director of his department announced the discovery of Element 93 before he could fully verify his findings. And he recalled a paper by a German chemist, Ida Noddack, published in an obscure journal of chemistry. It argued that what Fermi thought was a new element may have been fragments of uranium nuclei. Fermi double checked his findings at the time and concluded that Noddack couldn't be correct.[9] Now he wondered if she was right. Had he split the uranium nucleus, but didn't know it?

CHAPTER FOUR

NUCLEAR FISSION

In the years following Fermi's 1934 announcement of a new atom, a German chemist Otto Hahn, his young associate Fritz Strassmann, and their colleague Lise Meitner, an Austrian physicist, continued Fermi's neutron bombardment experiments looking for transuranic elements heavier than Fermi's Element 93 and getting promising results. Meanwhile, in Paris, Irène Joliot-Curie and a Yugoslav associate Pavel Savitch published the results of their own neutron bombardment experiments which contradicted those of Hahn. Hahn scoffed at their findings and suggested they issue a public retraction.[1]

Their disagreement was reignited later that year when Ida Noddack, the chemist who had first criticized Fermi's findings, gave a lecture before an international gathering at the Soviet Academy of Sciences in Leningrad. Her topic: methods for predicting chemical elements. Hahn was in the audience. He and Noddack had crossed swords earlier in 1934 when her paper critical of Fermi was published. Hahn called her reasoning absurd, an opinion shared by many other physicists. After the lecture, Hahn decided to settle the issue by repeating Fermi's experiment. With the help of Strassmann and Meitner, he began what he believed would be a relatively quick exercise retracing Fermi's steps. He was mistaken.[2]

It took several months, but they thought they had proven Fermi's claim only to discover an error that forced them to start over. Their quest would take four years and lead them to a place they never imagined. Fermi had discovered that neutron bombardment caused heavy atoms—those with the most protons—to add a proton to their nuclei through a process called beta decay. Platinum, with seventy-eight protons, for example, turned to gold with seventy-nine protons. Gold transformed into mercury with eighty protons. Scientists worldwide had demonstrated the transformation time and time again through multiple experiments. Fermi's theory of beta decay was proven science as irrefutable as Newton's theory of gravity or Galileo's theory that the earth and planets revolved around the sun. So thought Hahn and colleagues. But they were wrong.

This time, instead of gaining a proton, the uranium atoms shed four protons, transforming them into what appeared to be radium with eighty-eight protons and leaving Hahn totally baffled. Louis Pasteur, the French biochemist, famously commented that "discovery comes to the prepared mind." Nothing in Hahn's extensive career had prepared him for this. He had fallen down a rabbit hole into a mysterious mirror-image, upside-down, black-is-white world where the old laws of chemistry didn't seem to apply. Hahn and colleagues conducted test after test. They exhausted every tool in their chemist's toolkit and published fourteen research papers in their quest to understand why the bombardment of uranium produced atoms with *fewer* protons instead of more, but the surprises kept coming.

After a particularly complex series of experiments, they discovered what they had thought was radium was, in fact, barium with fifty-six protons. Even more surprising, the barium was radioactive, suggesting it might have been part of the uranium atom's nucleus.[3] But that was impossible. Scientists using atom-smashing cyclotrons had fired charged particles packing nine million volts of energy at uranium atoms and failed to do any damage.[4] The barium nucleus was more than half the size of the uranium nucleus. The suggestion that a slow-moving neutron could break off a piece of the uranium nucleus that large was beyond absurd. It would be as if a tiny pebble lobbed at a large boulder had caused it to split apart.

Unable to explain what had happened, Hahn wrote his physicist colleague Lise Meitner, asking if she could "suggest some fantastic explanation."[5] An Austrian Jew, Meitner had fled Germany in the summer of 1938 after Hitler's annexation of Austria made her subject to the anti-Semitic laws targeting German Jews, and she eventually ended up in Stockholm. It was she and her nephew Otto Frisch who finally unraveled the mystery. In the atom's nucleus, positively charged protons repelled each other. They pushed, they shoved, and they battled, trying to get away from each other, but never could. That was because something called the "strong force" held them captive. Fermi's snail-paced neutrons had apparently disrupted that force.

Meitner was on Christmas vacation, skiing with her nephew, when the answer came to her. Suddenly, she understood what her German colleagues had not. A tiny, slow-moving neutron had very little energy. But, like the straw that broke the camel's back, it was just enough to tip the balance. The additional muscle it brought to the fight enabled the protons in the uranium nucleus to overpower the restraining strong force. In a process later called nuclear fission, the nucleus split, expelled two of its neutrons, and along with them, a burst of atomic energy.

The energy generated by each split uranium nucleus was infinitesimal, only enough "to make a visible grain of sand visibly [jump]."[6] But multiply that energy by trillions and trillions of split nuclei, and the total would become staggering. According to one estimate, the energy from a mass the size of a single raisin would be sufficient to light the lights, power the subways, and meet most of New York City's energy needs for a day.[7] But, as H. G. Wells apocalyptic novel envisioned, it could also be used to destroy that city.

On the night in 1934 when Fermi and students met to write a report on their experimental findings, no one imagined that an act of sheer whimsy, creating those slowpoke neutrons, would unlock "nature's greatest secret," opening the door to such exciting, but also terrifying, possibilities.

CHAPTER FIVE

NEWS OF NUCLEAR FISSION ROCKS SCIENTIFIC WORLD

Fermi acknowledged the possibility that his discovery of Element 93 may have been a mistake in a footnote to his not-yet-printed Nobel acceptance speech: "The discovery by Hahn and Strassmann," he wrote, "makes it necessary to reexamine all the problems of transuranic elements, as many of them might be found to be products of a splitting of uranium."[1]

In the world of physics, there were two kinds of scientists: the theoreticians like Einstein who used their fertile imaginations to explain natural phenomena, and the experimentalists who searched for answers by gathering and analyzing data through laboratory experiments. Fermi excelled at both, a rare combination.

Back in Rome, his students, awed by his brilliance, had nicknamed him "Il Papa"—the Pope—because they believed that he, just like the head of the Catholic Church, was infallible. But the atomic nucleus was a strange new world that scientists had only begun to explore. And, this time, "Il Papa" had lost his way.

The news of Hahn and Strassmann's historic breakthrough rocked the scientific world. "The most sensational discovery in modern physics since the discovery of radioactivity," proclaimed one observer.[2] Distinguished scientists in three countries had performed Fermi's neutron bombardment experiment, and none of them had understood what happened. "Everyone kicked themselves," said atom bomb historian Richard Rhodes. Rhodes interviewed University of California (U.C.), Berkeley physicist Glenn Seaborg, who had also missed the historic discovery. "And he told me, 'I spent the next three days in a blue funk. I had had this monumental discovery, this Nobel Prize level discovery in my hands, and I hadn't understood it.'"[3]

There was no putting the atomic genie back in the bottle. Each new experiment brought new evidence that split uranium nuclei emitted neutrons. In theory, those neutrons could go on to split more nuclei, yielding more neutrons, triggering a chain of events that would produce more and more atomic energy.

In late January 1939, Fermi and colleagues at Columbia University produced a chain reaction that generated what the *New York Times* called "the greatest amount of atomic energy so far liberated by man on earth."[4] But the research only hinted at the atom's true potential. Once dismissed as the stuff of science fiction, an atom bomb was now—theoretically, at least—within reach. The possibilities were exciting—and terrifying. The earth's atmosphere was 78 percent nitrogen. Would an atom bomb ignite the atmosphere? Would it turn the hydrogen in the oceans ablaze? Oppenheimer's colleagues didn't think so, but this was all new territory, and Manhattan Project physicist Emilio Segrè was worried. In an interview for the Atomic Heritage Foundation, he told historian Richard Rhodes, "People had calculated all . . . the hypotheses, possible hypotheses . . . and there was no chance of igniting the atmosphere. But I'm enough of a physicist to know that you calculate everything, and then something happens that you never dreamed."[5]

Despite those concerns, physicists throughout the United States and Europe raced to perform their own experiments. In France, Frédéric Joliot-Curie and colleagues published two scientific papers concluding that there were a sufficient number of neutrons freed in the nuclear fission of uranium to sustain a nuclear chain reaction. In Hamburg, Germany, a young physicist Paul Harteck read the Joliot-Curie papers and wrote the German War Office, alerting them to the prospects of a powerful new explosive based on nuclear physics. His letter warned that the first country that made the atom bomb would have a great advantage over the others.[6]

Meanwhile, Adolf Hitler's troops crossed into Czechoslovakia unopposed, bringing Europe to the brink of another world war. Terrified by the possibility of Hitler armed with an atom bomb, Fermi and colleagues, many of whom had fled Nazi anti-Semitism, attempted to warn Washington.

George Pegram, head of Columbia's physics department, was able to pull strings and arrange a meeting with an assistant to the chief of naval operations. Fermi had business in Washington the next day and was chosen to deliver the warning.

Fermi was sitting outside the admiral's office waiting to be ushered in when he heard the desk officer announce, "There's a wop outside." Not an auspicious start. Fermi did his best to explain the seriousness of the threat, but his low-key manner and halting English failed to convey the necessary sense of urgency. The admiral offered his thanks, and that was the last Fermi and colleagues heard of it.[7]

In July 1939, Fermi and Leo Szilard, a brilliant but irascible Hungarian physicist, put together an ambitious proposal to build a prototype nuclear reactor that would trigger a nuclear chain reaction and produce atomic energy. While Fermi prepared to go to the annual Summer School of Theoretical Physics in Ann Arbor, Michigan, Szilard set about trying to raise $35,000 to pay for the reactor. He first approached Pegram at Columbia. It was a lot of money, and Fermi would be in Ann Arbor for the next few weeks, so Pegram decided to put off any decision until the fall. Szilard approached the U.S. military looking for funding but failed to stir any interest from either the army or the navy.[8] Frustrated, he turned to Albert Einstein and convinced him to sign a letter he had written to President Roosevelt, alerting him that the Germans might be working on an atomic bomb and urging him to launch a similar program.

CHAPTER SIX

BOHR'S DISCOVERY

Fermi suspected that Heisenberg and the Germans were working on a nuclear reactor, but the science of nuclear fission was still rudimentary, and he took comfort knowing that harnessing atomic energy would likely take years to accomplish. Thus, when the Danish physicist Niels Bohr, then a visiting professor at Princeton University, discovered a seemingly insurmountable obstacle, Fermi and colleagues all breathed a sigh of relief.

Bohr was at breakfast in the Princeton faculty club when the conversation with his colleagues turned to nuclear fission. Once again, the scientists puzzled over the long-unsolved mystery—why the speed of the neutrons seemed so critical. Why did slow-moving neutrons split the atom's nucleus and release a burst of atomic energy and not fast ones? Suddenly, Bohr jumped from his seat and, trailed by a young associate, raced to his office across campus.

Breaking a piece of chalk in his rush to get his thoughts down on the blackboard, he drew diagrams representing two uranium atoms: Uranium 235 (U235) and the slightly heavier Uranium 238 (U238). Taking a moment to admire his work, he then shared with his young colleague the epiphany that had triggered his excitement. Neutron bombardment did not split every uranium nucleus. It only split the U235 nucleus. U235 was nature's atomic energy storehouse. The other atom, U238, was, in effect, its security guard. For every storehouse, there were 139 guards standing sentry, capturing neutrons, absorbing them, and preventing them from gaining access to nature's U235 source.[1]

Under normal conditions, the defenses were impenetrable, but nature's grand design had one flaw. Nature had programmed Uranium 238 atoms to capture intermediate and fast-moving neutrons, but not slow ones. Low-energy slowpokes could sneak past the guards and break into the nuclear energy storehouses undetected. Fermi didn't know it in 1934, but he had stumbled upon a back door, a secret passage into those storehouses, open only to slowpokes.

In 1939, there was no practical way to separate Uranium 235 atoms from U238, at least not on the mammoth scale required. As long as the scarce U235 atoms remained buried among all those U238 atoms, there was no chance of anyone harnessing enough energy to fuel a bomb.

Even if a way could be devised, conventional wisdom held that it would take decades to harvest enough U235 atoms for a single bomb. One expert estimated that "over 191 years would [roll by] before a single gram of U235 was obtained, and 75,000 years for a single pound."[2] "It can never be done," Bohr told his colleagues, "unless you turn the United States into one huge factory."[3]

H. G. Wells's vision of a world shattered by "crimson conflagrations of atomic bombs" was still just science fiction. Across the Atlantic, British Prime Minister Winston Churchill assured his secretary of state for air, "the fear that this new discovery has provided the Nazis with some sinister, new, secret explosive with which to destroy their enemies is clearly without foundation."[4]

But, then again, if any nation had the scientific know-how and the industrial capacity to overcome all the obstacles, it was Nazi Germany. In the 1920s and '30s, Germany was the indisputable world leader in physics, chemistry, and mathematics. German scientists had won one-third of the world's Nobel Prizes in science.[5] Germany's chemical and armaments industries were second to none. Germany had access to the latest nuclear fission research from British, Soviet, Italian, American, and French scientists, all of whom had openly published their work before the war. Nazi Germany controlled a large uranium mine in occupied Czechoslovakia.

When General Groves and Oppenheimer began work on the Manhattan Project, they had very little hard intelligence concerning Germany's atom bomb program. They didn't have a network of spies or the sophisticated electronic sensors and satellites that are common today. Most of their information came from individuals who had fled Nazism. But the Allied scientists did know the man leading Germany's atomic research program: a man universally recognized as one of the most brilliant theoretical physicists in the world. They had heard him lecture at international conferences and read his research papers. He was Germany's youngest full professor at age twenty-six and a Nobel Prize winner at thirty-three: Werner Heisenberg.

CHAPTER SEVEN

HEISENBERG IN MICHIGAN

When Heisenberg traveled to the United States in the summer of 1939, it was a victory lap of sorts. He had just come through a frightening year-long investigation by the Waffen-SS, the armed wing of the Nazi Party, during which he was called to Schutzstaffel (SS) headquarters and interrogated and his house bugged. Images of brown-shirted Gestapo stormtroopers coming to arrest him haunted his dreams for years.[1]

Werner Heisenberg. SUEDDEUTSCHE ZEITUNG PHOTO / ALAMY STOCK PHOTO.

After watching the Nazis gut German universities of their best Jewish professors and replace them with second-rate party loyalists and then ban the teachings of Albert Einstein, Heisenberg had felt compelled to step in. He attempted to persuade Nazi officials to rescind the dismissals while also trying to convince the fired professors to remain in Germany.[2] Risking his career and his life, he published articles in the Nazi Party and SS newspapers defending the scholarship of Einstein and Niels Bohr, two of the world's most distinguished theoretical physicists, and both Jewish. Johannes Stark, a German Nobel laureate and Nazi loyalist, responded with a vitriolic attack in which he labeled Heisenberg, who was Christian, a "white Jew." That attack triggered the SS investigation.

Fortunately for Heisenberg, he had friends in high places. His father had taught high school and been friendly with the father of Heinrich Himmler, commander of the SS. Heisenberg's mother wrote Himmler's mother, and, after many months of investigation, Heisenberg finally received a letter from the SS commander, stating, "I take pleasure in being able to inform you . . . there will be no further attacks on your person."[3] Heisenberg had dodged a bullet and struck a blow for German science. He could continue teaching the theories of Jewish scientists. He could teach the theories of relativity and quantum mechanics, but Himmler imposed one condition: Heisenberg was forbidden to credit the Jewish scientists responsible for those advances.

Heisenberg had been a strong defender of what the Nazis called "Jewish science" ever since 1933 when Hitler came to power. But it was the glory of German physics that motivated him far more than the well-being of his Jewish colleagues. The dismissal of his former professor Max Born, a Jew, in 1934 made that clear.

Born responded to the dismissal by leaving Germany and moving his family to Cambridge, England. Heisenberg subsequently intervened with Nazi authorities and succeeded in winning a compromise. Born would be allowed to return, but he would not be allowed to teach. He could only do research. When Heisenberg informed Born about the decision, Born inquired about his family, and Heisenberg told him that his family was not included in the invitation. Born turned down the invitation and didn't return to Germany until after the war. He and his wife went back to retrieve some items they had left behind, and they paid a visit to Heisenberg. An associate of Born describes what happened. "They were met with anti-Jewish sneers and obscenities"—the associate doesn't make clear who behaved that way—"and in the end Heisenberg spat on the floor at Max Born's feet." Born's wife finishes the story: "And my poor Max wept."[4]

In 1939, Heisenberg won immediate approval for his application to travel and lecture in the United States. By the time he arrived in the States, hopes of avoiding another world war had all but vanished. The specter of one of history's greatest villains armed with an atom bomb haunted America's theoretical physicists, but if Heisenberg was troubled, he didn't let on. An avid outdoorsman, he enjoyed climbing the Rockies and exploring the Grand Canyon, Yellowstone, and the Sierras.[5] At U.C. Berkeley, he seemed totally at ease while lecturing at Oppenheimer's Wednesday afternoon seminar. Afterwards, over tea, he went on at length talking about San Francisco cable cars and the technology that enabled tracks and power lines to cross.[6]

To Oppenheimer, Heisenberg's behavior seemed calculated. To him, it was all an act. Writing in 1980 after her husband's death, Elisabeth Heisenberg insisted it was genuine: "At the time, he still firmly believed in the 'world community of physicists' he spoke about so often." Long isolated in Nazi Germany, he was happy to be back in that world. "Never having identified himself with the policies of the Nazis," Elisabeth Heisenberg wrote, "he was of the firm conviction that old friendships could outlast political differences."[7]

At U.C. Berkeley, Robert Oppenheimer warned an associate that Heisenberg could not be trusted. Oppenheimer was Jewish and had relatives in Germany, so his stance was not surprising. But his animus was likely motivated by more than anti-Semitism. The brilliant young physicist Oppenheimer had met in Göttingen had won a Nobel Prize; Oppenheimer had not, and he didn't like being second best.[8]

Heisenberg's work in quantum mechanics was revolutionary. It had helped develop a new understanding of the atom and the behavior of electrons. Under the old laws of classical physics, electrons were thought to orbit the nucleus just like the earth revolved around the sun; their locations and velocities were thought to be predictable just like that of the earth and the planets. Not so, said twenty-five-year-old Heisenberg. The laws of classical physics did not apply to small objects like electrons. According to his famous Uncertainty Principle, it was impossible to determine both the exact position and exact velocity of an electron at any given moment. The more precise one measurement, the less precise the other. The best one could do was compute the probability of the electron being at a specific location or its specific velocity.[9]

The world's most famous physicist Albert Einstein disagreed that quantum mechanics could be that uncertain. "God doesn't play dice with the universe," he argued.[10] Nevertheless, Heisenberg's Uncertainty Principle became a pillar of modern quantum mechanics.

Oppenheimer was also brilliant—the West Coast doyen of American theoretical physicists. He had done important work in the fields of nuclear physics, astrophysics, and spectroscopy,[11] but his star had not risen to the heights that Heisenberg's had, and he didn't have Heisenberg's global stature—a reality that weighed heavily on him.

Oppenheimer and Heisenberg met a second time that summer of 1939, at the University of Chicago, where Heisenberg gave a lecture on cosmic rays. The two had competing theories on the subject, and after the lecture, Oppenheimer confronted Heisenberg. Heisenberg, like Oppenheimer, had a healthy ego. Heisenberg was arrogant, extremely competitive, and did not like being challenged. Once when a visiting Japanese physicist beat him at table tennis, a sport he worked hard and excelled at, he didn't show his face at the office for over a week.[12] When Oppenheimer dared to confront him, tempers flared, and they got into a shouting match.

After Chicago, Heisenberg boarded a train and traveled east to Purdue University in Indianapolis where he and his friend Hans Bethe, a German-born Cornell University physicist, had been invited to lecture. Bethe would later be named head of the Manhattan Project's all-important Theoretical Division.[13] From Indiana, Heisenberg went on to Ann Arbor and the University of Michigan for another series of lectures. The annual summer symposium in theoretical physics offered Americans an opportunity to learn from their European colleagues. In 1939, the impact of Nazi anti-Semitism and the flight of Jewish scientists had not yet fully registered, and Germany was still considered the world leader in physics and chemistry.

In Ann Arbor he met up with several old friends, including Enrico Fermi. Fermi had time on his hands that summer. His nuclear fission research had stalled for lack of funds.[14] The University of Michigan summer school was always a stimulating and festive gathering and offered a welcomed opportunity to catch up with colleagues from abroad.

Heisenberg had based his Uncertainty Principle, in part, on Fermi's work. The two had been friends for more than a decade, so Fermi had license to press him on the issue, worrying them all. "There is now a real chance that atom bombs may be built," he told Heisenberg.[15] If war came, would he support the German cause? Would he build Adolf Hitler an atom bomb?

Heisenberg dodged the question. He didn't express any moral or political objections to arming Hitler. Instead, he said, "atomic developments will be rather slow . . . I believe that the war will be over long before the first atom bomb is built."[16]

While attending the Michigan seminars, Heisenberg always stayed with his longtime friend, the event's principal organizer, Sam Goudsmit and his family. Goudsmit was a genial man popular with his colleagues as well as his students, whom he invited to his home on weekends for pancake breakfasts. "Sam's rather peculiar for a physicist," commented one friend. "Most of us are just as sensitive as he is, but to inorganic matter rather than to people and their private worries."[17]

Everywhere Heisenberg went, his American friends urged him to bring his family to the United States and to accept one of the many university professorships offered him, but Heisenberg refused. Win or lose, Germany would need leaders after the war to plot a path forward, and, in typical fashion, he saw himself as one of those leaders.[18]

Heisenberg had made his decision before coming to the United States. In his view, incompetent Nazi bureaucrats were undermining the quality of theoretical physics education in Germany. He sympathized with Hitler's ambition of returning Germany to a position of prominence in Europe but opposed what he called his "excesses."[19] He objected to anti-Semitism and Nazism—not because they were bad or because he was altruistic, but because they were bad for German science. Protecting and promoting German science was at the core of his being.[20]

Faced with a choice of escaping the coming war or staying to defend his disparaged science, Heisenberg put science and the fatherland above his personal safety and that of his family. He bought a cabin for his wife and five young children in the mountains of southern Germany, hoping they would be safe there during the war, while he stayed behind at the university, tending to scientific matters. To him, having lived through the turmoil of hyperinflation and the Great Depression following World War I, Nazism was merely a passing storm. He told Goudsmit that the Nazi regime would eventually fail and that he and others like him would then step in.[21] He hoped his American friends would understand.

From Michigan, Heisenberg traveled to New York City where he lectured at Columbia University, visited his aunt and uncle in the suburbs, then boarded the all-but-empty ocean liner *Europa* for the voyage home.[22] In his luggage he carried photographs of his trip, mementos of happier times. The trip would take four or five days, plenty of time to reflect on what lay ahead. As he wandered alone on the deck peering out at the ocean, he wrestled with the question his visit had left hanging. Heisenberg was a patriot. He loved his country but had only contempt for the Nazis. What would he do? Was he duty bound to build Adolf Hitler an atom bomb? Time to decide was running out.

CHAPTER EIGHT

LIGHTNING WARFARE

> **WAR DEPARTMENT FILM BULLETIN**
>
> Poland, September 1939, the German Wehrmacht begins its ruthless march of conquest and sets the stage for World War II. Poland's 34 million inhabitants crushed, scattered and enslaved.

Rejecting the trench warfare and static defenses of the First World War, the German war machine adopted a strategy it called "blitzkrieg"—lightning warfare. After smashing through Poland's border defenses, the first wave, made up of German tanks supported by Luftwaffe bombers and fighters, encircled Polish troops. The second wave of motorized infantry followed, securing their gains. Then the regular infantry took over, releasing the motorized infantry to race after the tanks that had already rolled deeper into Polish territory. It took the Wehrmacht just three weeks to defeat Poland's smaller and poorly equipped forces.

In a preview of the atrocities to come, Nazi paramilitary death squads executed 20,000 Poles, most of them Jews and intellectuals, and buried them in mass graves. The Germans bombed the city of Warsaw and strafed civilian refugees as they clogged the roads trying to escape. In all, more than 150,000 Polish civilians were killed.[1] Two days after the invasion, England and France sided with Poland, launching what would become the bloodiest war—with the highest civilian death toll—in history.

To memorialize his army's stunning victory, the führer traveled to the Free City of Danzig where he delivered a speech in which he extended an olive branch clenched in an iron fist. He vowed that Germany had "no war aims against Britain and France," but at the same time threatened any nation that dared oppose it, saying, "We will employ a weapon with which we could not be attacked." Hitler's threat set off alarms throughout the continent and filled newspapers with speculation about a secret weapon, a Nazi "death ray."[2]

A short while earlier, Germany's Army Ordnance Research Department in Berlin had arranged for draft notices to be sent to eight scientists for a special assignment: to discover whether it was possible to harness atomic energy to make a bomb or generate electricity.[3] The men were all experimental physicists. Werner Heisenberg, a theoretical physicist, was not among them.[4]

The head of the department was a loyal Nazi, thirty-four-year-old Kurt Diebner. An experimental physicist, he believed experimentalists were better suited than theoreticians for the task ahead. For their part, theoreticians considered themselves intellectually superior to the experimentalist, resulting in an unspoken tension between the two. Diebner's relationship with Werner Heisenberg was especially strained. Heisenberg had once embarrassed him by calling his academic research "amateurish," and Diebner had not forgotten.[5] Heisenberg's supporters eventually prevailed on Diebner to change his mind and include the distinguished Nobel Prize winner in what came to be known as "the Uranium Club." But relations between the two men remained strained.

Diebner did not have the gilt-edged credentials of Heisenberg or the other Club members. He had a PhD in physics but was never able to obtain a professorship, so he took a job outside academia as advisor to the Army Ordnance Research Department where he evaluated literature on nuclear fission and tried desperately to get the army interested in the potential of nuclear energy. It was a losing battle until the director of the Hamburg Institute for Physical Chemistry sent the Army Ordnance a memorandum echoing what Diebner had been saying. At that point, Diebner's fortunes changed. He was made head of the newly founded nuclear research department and immediately requested military funding to set up a laboratory.[6]

The laboratory's first goal: to build a uranium nuclear reactor, an intermediate step that could lead to a uranium bomb. The reactor would demonstrate the viability of nuclear fission theory and help ensure continued government support for what would be a very expensive program with no guarantee of success.[7] Wasting no time, Diebner began parceling out research goals for the nine members of the Uranium Club and their institutes. He assigned responsibility for designing the nuclear reactor to the Kaiser Wilhelm Institute for Physics in Berlin under his direction and to Heisenberg and the Leipzig university physics institute.[8] The nine Uranium Club members were spread across Germany. The War Office proposed bringing them together in Berlin at the Kaiser Wilhelm Institute, but the members balked at the idea, preferring to remain in their own little fiefdoms.[9] As a result, Club members would at times end up competing among themselves for the limited supplies of uranium.[10]

At his university office in Leipzig, Werner Heisenberg hunkered down and produced a report on applied nuclear fission theory that focused on the construction of an energy-producing machine but also enticed his Nazi overseers with the possibility of a super powerful bomb. He was confident that he could make significant progress on the energy machine before the end of the war, but the explosive would not likely be available until sometime in the distant future. Heisenberg's report would become the blueprint for Germany's atomic research program. He would conduct the reactor experiments in Leipzig and also hold sway over those in Berlin. Heisenberg was not the director of Germany's atomic research program, but his was certainly the dominant voice.[11]

After years of risking his freedom and his life battling Nazi critics who disparaged the work of Jewish physicists, Heisenberg saw an opportunity to prove the value of theoretical physics and at the same time win himself and his young colleagues deferments from being drafted into the army. Instead of serving in the army, he would build the führer a "uranium machine," a nuclear reactor.

The Germans had a slogan, "We must make use of physics for warfare." Heisenberg turned the slogan around: "We must make use of warfare for physics!"[12] In a postwar memoir, Elisabeth Heisenberg wrote that she asked her husband what he would do if he were ordered to build an atom bomb. He assured her, as he had Fermi before the war, that would never happen. "Don't worry," he told her. "We will not build atomic bombs. The development of a bomb is a gigantic project. It would probably take years."[13]

"INVASION OF DENMARK & NORWAY," BBC RADIO, APRIL 9, 1940, ALVAR LIDDEL REPORTING

It was about dawn this morning when the first reports came in saying that German troops are crossing the frontier into Denmark. At the same time attacks are being delivered from the sea on a number of Norway's biggest ports.

Germany, France, and England refrained from large-scale military operations after the initial invasion of Poland. But that period of relative calm, known as "the Phony War," lasted only eight months.

As Hitler's armies rolled across Europe, a group of physics professors at the University of Hamburg wrote the Reich Ministry of War warning that the latest atomic research held out the possibility of creating explosives far more powerful than conventional weapons and "in America and in other Anglo-Saxon countries as well as in France pure atomic physics is presently much more intensively pursued than here."[14]

Meanwhile, the uranium mine in the Czech Sudetenland, now under German control, stopped exporting uranium ore. In May 1940, German troops invaded Belgium and seized the town of Olen where the Belgian mining company Union Minière du Haut Katanga had its main refinery.[15] The world's largest source of uranium, the company had stockpiled thousands of tons of uranium there.[16] That same month, the *New York Times*, citing highly reliable channels, reported that the Nazi government was experimenting with a nuclear reactor at a research laboratory outside Berlin. "Every German scientist in this field—physicists, chemists and engineers," it stated, "have been ordered to drop all other research and devote themselves to this work alone."[17]

Also in May, German troops seized a laboratory at the College de France in Paris where Frédéric Joliot-Curie was building a cyclotron. A cyclotron, or "atom smasher" as it was known, was essential equipment for designing an atom bomb. The Kaiser Wilhelm Institute of Physics (KWIP) had contracted to build one in 1938, but as the nation prepared for war, the necessary labor and materials went to higher priority projects and construction was delayed.

Kurt Diebner, his boss Erich Schumann, and their translator Wolfgang Gentner rushed to Joliot-Curie's lab shortly after the Germans marched into Paris. There, Schumann ushered Joliot-Curie into an office away from his staff and informed him that his cyclotron would soon be dismantled and shipped to Berlin.

As Schumann was leaving, Gentner, his translator, hung behind and whispered to Joliot-Curie to meet him at a nearby café that evening. Unbeknownst to Schumann, Gentner had worked in Joliot-Curie's lab before the war and was opposed to the Nazi regime.[18]

At their meeting, Gentner revealed the details of a plan he would propose to Schumann: that the cyclotron stay in Paris and that Joliot-Curie allow German physicists to use it. Joliot-Curie initially refused but finally agreed under two conditions: the Germans could only do peaceful basic research—nothing geared specifically to building an atom bomb—and Gentner had to be in charge of the lab. Gentner had worked with Ernest Lawrence, who had won a Nobel Prize for his invention of the cyclotron, in California before the war, so he would be invaluable to the lab.[19]

Gentner took the compromise to Schumann, and he agreed. All during the war, Gentner and Joliot-Curie did their best to make sure that research that could help Heisenberg build a bomb did not reach Berlin.[20]

The warning signs that Hitler intended to build an atom bomb were flashing red. British intelligence was worried. But with the Luftwaffe pounding Britain's coastal defenses in preparation for an invasion, they had a more imminent threat to worry about. In 1940, conventional wisdom held that because U235 atoms were so rare, a critical mass with sufficient U235 atoms to sustain a nuclear chain reaction would require many tons of uranium. The bomb would be far too heavy to carry in an airplane. So said the conventional wisdom.

But what if the conventional wisdom was wrong?

Early in 1940, two Jewish physicists, Lise Meitner's nephew Otto Frisch, an Austrian, and Rudolf Peierls, a German, then working at Birmingham University in England, began pondering: What if Heisenberg and colleagues could strip away all the U238 atoms that captured speeding neutrons? Without them, there would be nothing to stop the neutrons from splitting U235 atoms and triggering a chain reaction.

To answer their question, they began experimenting with different ways to separate out the U238 atoms. In the process, they ran the numbers and were shocked by what they discovered. If the U238 atoms were removed, the critical mass necessary to build a bomb was no longer measured in tons. It would be just a few pounds! "At that point," said Frisch, "we stared at each other and realized that an atomic bomb might after all be possible."[21]

Following their discovery, conventional wisdom did an about-face. The Maud Committee, a committee of British scientists set up to study nuclear fission, concluded that a U235 bomb was "practicable and likely to lead to decisive results in the war."[22] The committee recommended the Allies give developing a bomb the highest priority.[23]

CHAPTER NINE

A SECOND ROUTE TO AN ATOMIC BOMB

The Maud committee report took many months to make its way across the Atlantic. When Lyman Briggs, the chairman of President Roosevelt's newly created Uranium Committee, got his copy, he tucked it away in his safe instead of distributing it. The research program the Maud report recommended would have cost many millions of dollars. Worried about the impact on his reputation should it fail, Briggs elected not to show the document to his fellow committee members.[1]

On a July morning in 1940, Carl Friedrich von Weizsäcker, a young colleague of Werner Heisenberg, was reading a copy of an old scientific journal on his subway ride to the Kaiser Wilhelm Institute for Physics in Berlin when his mind began to wander. He and his colleagues had been so focused on nuclear fission and the role of U235 that they had paid little attention to the common, but non-fissionable, U238 atom. What became of it after its nucleus absorbed a neutron? He knew it wouldn't release a burst of atomic energy. But what if it decayed and gave birth to a daughter atom? What if the daughter was fissionable? U238 atoms were so common, the odds of building an atom bomb would skyrocket. U238's daughter would become an atomic energy bonanza. Germany wouldn't need to invent new separation technologies or build massive plants to physically separate U238 from U235 atoms. The new atom, unlike U235, would be separable by a commonly used chemical process. Von Weizsäcker couldn't wait to get off the train and alert the War Office.

Von Weizsäcker didn't know it at the time, but a month earlier, two American physicists at U.C. Berkeley had traveled the same road, but they did more than theorize. Edwin McMillan and Philip Abelson used the university's cyclotron to bombard a sample of uranium, and, sure enough, its U238 atoms decayed into daughter atoms. But the daughters turned out to be unstable, decaying further. Was that next generation of atoms fissionable? Finding the answer would involve a long, multi-step procedure, and McMillan and Abelson had other commitments. McMillan had contracted to go to the Massachusetts Institute of Technology (MIT) to work on a more urgent national security matter—the development of radar—and Abelson, who had worked with McMillan only briefly while vacationing in Berkeley, had to return home to his full-time job. As a result, before leaving Berkeley in May 1940, they mailed their findings to a publisher, leaving the question of a fissionable daughter atom unanswered.[2]

Their report appeared in the June 15, 1940, edition of the scientific journal *Physical Review*. When James Chadwick, a distinguished British physicist who had won a Nobel Prize in 1935 for his discovery of the neutron, received his copy, he was livid. France, with the largest army in Europe, had been the one real hope to slow the German advance, but it had collapsed in just thirty-five days. Now the Germans were just across the English Channel twenty miles away preparing to invade England. Britain's survival was on the line, and American scientists had just alerted the Germans to a more efficient way to build an atom bomb.

At Chadwick's urging, the British Embassy in Washington sent an attaché to U.C. Berkeley to register an official protest.[3] Meanwhile, in Berlin, Von Weizsäcker read the McMillan–Abelson paper and realized that he had bet on the wrong generation of atoms. He quickly fired off a new memo to the War Ordnance Office correcting the mistake.

Eight months later, Glenn Seaborg, a brilliant twenty-six-year-old chemist at U.C. Berkeley, asked McMillan and Abelson for permission to pursue the experiments they had begun. After weeks of work, he and his coworkers discovered Pu239, the elusive fissionable atom McMillan, Abelson, and Weizsäcker had been searching for. Seaborg named it Plutonium. With that discovery, Seaborg opened a second route to an atom bomb, a seemingly less complicated, more direct route.

Fritz Houtermans, a physicist working on chain reaction theory at a private laboratory in Berlin, made the same discovery.[4] Horrified by the destructive potential of a plutonium-fueled bomb, Houtermans contacted his old friend Werner Heisenberg and was surprised to learn that he knew about the new atom and was working to develop a nuclear reactor. Houtermans knew Heisenberg was not a Nazi. So why was he working on a nuclear reactor? Was he planning to build Hitler an atom bomb? Heisenberg never answered that question.

In March 1941, Houtermans paid a visit to Fritz Reiche a young Jewish physicist who had been dismissed from his university job due to Nazi racial laws and had spent six years struggling to find work. Houtermans knew he and his family were preparing to emigrate to the United States, and he asked Reiche to carry a message for him. It was very sensitive, and Reiche, his wife, and his daughter would undoubtedly be imprisoned or perhaps killed if he were caught, so Houtermans insisted he memorize it.

A friend met Reiche and his family at the dock in New York and drove them to Princeton University where they stayed with a physicist friend. Upon learning about Reiche's message, he arranged a dinner with about a dozen, mostly refugee, scientists. Speaking from memory, Reiche told the dinner guests that Heisenberg was "trying here, hard . . . to hinder the idea of making a bomb. But . . . will not be able to withstand longer the pressure from the government."[5]

Houtermans didn't work with Heisenberg and was not in a position to know exactly what he was trying to do. All one can say with certainty is that he *believed* Heisenberg was trying to obstruct the effort. The historical record is often contradictory. Author Robert Jungk interviewed several of Heisenberg's colleagues for his 1956 book *Brighter Than a Thousand Suns*. According to Jungk, Heisenberg, Weizsäcker, Houtermans, and "at least ten other prominent German physicists" agreed that they must "keep the departments in the dark about the now imminent feasibility of manufacturing such bombs."[6] Journalist Thomas Powers, however, writes in his book *Heisenberg's War: The Secret History of the German Bomb*, published in 1993, that Jungk later had a change of heart. He felt he had been misled by the Uranium Club physicists and regretted "spreading the myth of passive resistance."[7]

CHAPTER TEN

THE BATTLE OF BRITAIN

BBC RADIO BROADCAST, JUNE 18, 1940, WINSTON CHURCHILL

The Battle of France is over. The Battle of Britain is about to begin. Upon this battle depends the survival of Christian civilization . . . if we fail, then the whole world including the United States . . . will sink in the abyss of a new dark age . . . Let us therefore brace ourselves to our duty, so . . . that if the British Empire and its commonwealth last for a thousand years, they will still say "this was their finest hour."

On July 10, 1940, three weeks after Churchill's broadcast, the Battle of Britain began. It was the first battle in history fought solely in the air. Hundreds of German bombers targeted Britain's coastal defenses and channel, shipping in daylight raids while Luftwaffe and British Royal Air Force fighters battled in the skies above. Hitler hoped to defeat the RAF and force the British to accept a negotiated peace. But heavy losses eventually drove the Luftwaffe to abandon their daylight raids and attack at night instead.

As a rule, nighttime bombing raids were extraordinarily inaccurate. An after-action report conducted later in the war found that only 5 percent of the RAF's bombs landed within a mile of their targets.[1] Given their own dismal record, British Bomber Command was shocked when a nighttime raid by twenty Luftwaffe bombers launched from airfields more than two hundred miles away scored eleven direct hits on a Spitfire aircraft factory near Birmingham.[2] The Spitfire was critical to Britain's control of the skies. Without the factories, Britain couldn't churn out the numbers of fighter aircraft necessary to win the air war. Without the Spitfires, Britain couldn't survive.

The Germans were brilliant engineers, but they had made a costly mistake. By trying out their new technology in England three weeks before the launch of their principal air campaign, they had triggered a rush by British intelligence to reexamine evidence gathered from downed German bombers and their crews, including a scrap of paper on which the plane's navigator had written "radio beacon Knickebein from 0600 hours on 315" and the transcript of an interrogation with a German prisoner of war (POW) who mentioned that Knickebein was a shortwave radio signal. Those led investigators to reexamine the radios on German bombers and to discover that they received a wider range of frequencies than comparable RAF radios. Those discoveries pointed to a possible answer to a question British intelligence analysts had been pondering for some time: Why had the Germans built several large radio transmitters along their border with France, and why were they aimed at England?[3]

R. V. Jones, a young physicist working for the Air Ministry, theorized that the Germans had beamed a radio signal at the Spitfire factory, and the bombers had followed it to their target.[4] Senior scientists dismissed the idea. Radio signals travel in a straight line. They didn't bend around the earth. They argued that a bomber flying at 20,000 feet could receive such a signal up to 180 miles away, but no farther. The spitfire plant was 260 miles from the closest transmitter.[5]

Outranked but determined, Jones arranged for an aircraft equipped with a radio receiver similar to that of the Germans to fly the route the Luftwaffe had flown, and, sure enough, it discovered a secret radio signal aimed at the factory. Masters of the impossible, the Germans had built their transmitters on high ground so that planes flying at twenty thousand feet could receive their transmissions throughout much of England. (Jones would later discover the system was more sophisticated than he had imagined. As the German bombers approached their targets, they encountered intersecting crossbeams from other transmitters that automatically triggered the release of their bombs).[6]

Armed with that intelligence, the British rushed to invent countermeasures, some to jam the radio signal and others to transmit false signals luring German pilots away from their intended targets. In response, the Germans changed their frequencies and built new transmitters. The "Battle of the Beams," as it came to be called, was the first skirmish in a secret war pitting German and Allied scientists, weapons engineers, intelligence analysts, and spies against each other—a war of cutting-edge science and technology that would pave the way for the race to build an atomic bomb.

Unfortunately, the counter measures took time to develop and were not available in sufficient numbers to fully protect British cities. For fifty-five days, Hitler's bombers terrorized the residents of London, Birmingham, Coventry, Southampton, and other cities with nighttime raids that came to be known as the Blitz.[7] The Blitz would ultimately claim forty-three thousand lives.

While British citizens huddled in subways and basement shelters trying to escape the bombs, Heisenberg spent much of his time on a train traveling the two and a half hours between his home and the university in Leipzig and Berlin where he and his colleagues were supervising construction of a new reactor laboratory.[8] To keep curious passersby away, they named the laboratory "Virus House."

At the same time, in the summer of 1940, Sam Goudsmit was teaching physics at Harvard in Cambridge, Massachusetts. Several of the university's physics professors had been called to work on U.S. Army and Navy research programs, and an acquaintance had telegrammed him asking if he would take a temporary job teaching summer school. After consulting his wife and daughter, Goudsmit got a leave of absence from Michigan and headed east.

While in Cambridge, he met physicists who were working at MIT on a secret program to develop airborne radar. Those contacts eventually led to an invitation from MIT for Goudsmit to come on board. It took several weeks for Goudsmit to get the necessary security clearance, much longer than he expected. He called MIT several times to ask why it was taking so long, but never got a straight answer. He didn't know it, but he was under surveillance that whole time. Military intelligence tapped his phone and trailed him every time he left the hotel to make sure he wasn't contacting foreign agents. His security clearance finally came through in early 1942, and he was fully engaged working as editor of secret documents and acting head of a group investigating the theory of radar.[9]

CHAPTER ELEVEN

ENRICO AND LAURA IN AMERICA

BBC RADIO, JUNE 22, 1941, WINSTON CHURCHILL

At 4 o'clock this morning Hitler attacked and invaded Russia. . . . Suddenly, without declaration of war . . . German bombs rained down from the sky upon the Russian cities.

The nightly bombing raids on England's cities ended unexpectedly in June 1941 when Hitler redirected his bombers to the east, double-crossing his erstwhile ally, Soviet Premier Joseph Stalin. It was the largest land invasion in military history. Its goal: not merely to capture territory, but to annihilate what Adolf Hitler claimed was a conspiracy of Jews and Bolsheviks (Communists) threatening all of Europe.

Enrico and Laura Fermi followed the latest developments from a furnished rental apartment in New York City overlooking the Hudson River where they began acclimating to life in the United States. Laura had never cooked before, but she bought a cookbook, translated the quantities to metric measures, and got started.[1] Enrico had been coming to the United States every summer for the past four years to attend the University of Michigan physics conferences, but his English still needed work. He knew that his colleagues and students would hesitate to correct the esteemed Professor Fermi, so he let them know that they would be awarded a penny for each correction.[2]

An ocean separated the Fermis from the war in Europe, but they still felt threatened. Ku Klux Klan rallies were increasing. Fascist ideology was becoming more acceptable with the likes of Charles Lindbergh praising Adolf Hitler as "undoubtedly a great man," and a widely popular "radio priest," Father Charles

Coughlin, delivering pro-Hitler, anti-Semitic sermons weekly to some thirty million listeners.[3] In January 1940, federal officials foiled a plot by members of the Christian Front, a right-wing group bent on raiding government offices, assassinating congressmen, "liquidating Jews," and setting up a government modeled after Hitler's dictatorship over Nazi Germany. Seventeen men were arrested, and a small arsenal of weapons seized.[4] Meanwhile, a growing number of Americans were joining the German American Bund, a pro-Nazi organization.

In February 1939, the Bund held a rally at Madison Square Garden in New York City, which drew twenty thousand supporters. Marching to the beat of snare drums, men in Nazi uniforms carrying American flags paraded into the garden where they were greeted by an enthusiastic crowd of supporters, their arms raised in a Nazi salute. Standing in front of a towering portrait of George Washington, Fritz Kuhn exhorted the crowd to "protect ourselves . . from the palsied hand of Jewish communism" while a newsreel camera recorded Nazi guards beating a lone man who dared to protest.[5]

WAR DEPARTMENT FILM
WHY WE FIGHT, THE NAZIS STRIKE

And this is Fritz Kuhn, leader of a German American Bund hiding behind an American flag but taking his orders from Berlin . . . and copying the methods of Berlin.

Laura Fermi would later write: "There seemed to be little doubt that Germany would gain a final victory in Europe and that such a victory might mean a Nazi domination of America . . . through increased power of the German Bunds."[6]

Worried that Nazism was taking root in the United States, the Fermis buried the remaining Nobel prize money in the basement of their new home in Leonia, New Jersey, for safekeeping and spent evenings with friends plotting their escape from the United States. They figured the Germans would control the Atlantic. So they planned to make their way by sailboat from Florida to an island in the Pacific. "We prepared to become modern Robinson Crusoes in some far away desert Island," wrote Laura Fermi. Enrico's "delight at the prospect of experimenting with compass and sextant was encouraging." Laura took charge of their clothing needs, making sure their island "colony would not go naked in years to come."[7]

CHAPTER TWELVE

HARNESSING ATOMIC ENERGY

In August 1941, while Laura debated whether to provision their boat with "cotton seed and spinning wheels or . . . bolts of cloth," Enrico, Leo Szilard, and colleagues launched their first attempt to build a full-scale nuclear reactor.[1] By then, Vannevar Bush, director of the Carnegie Institution, and other prominent scientists and engineers had successfully lobbied President Roosevelt to set up the National Defense Research Committee (NDRC) to fund civilian weapons development research. And Szilard had managed to secure $40,000 of its funds for their experimental atomic pile, the nuclear reactor.[2]

As Fermi and Heisenberg began their quest to harness the atom's energy, they faced a critical decision: choosing a moderator to slow down the neutrons emitted by the split nuclei. The newly liberated neutrons traveled at speeds of roughly ten million meters *per second*, far too fast to be absorbed by the uranium atoms. Graphite had a layered structure of carbon atoms that would not absorb the neutrons but would cause them to ricochet from atom to atom, slowing them down. The slower, tortoise-like neutrons would then move only slightly faster than the Uranium 235 atoms, improving the odds that U235 would absorb them as they passed by.[3] Graphite was inexpensive and readily available, so it was clearly a good choice.

But there was a problem. Tests conducted by Enrico Fermi's team at Columbia revealed impurities in the graphite that absorbed neutrons and would shut down any chain reaction. The problem was solved when Szilard went back to the National Carbon Company that had supplied the graphite, and they were able to purify it.

The United States had not yet entered the war, and Fermi was eager to share his graphite research with his physicist colleagues worldwide, so he prepared to publish the test results. When Szilard learned what he was planning, he and the normally unflappable Fermi got into a heated argument. American physicists had unwittingly helped Heisenberg once by publishing critical plutonium research; Szilard was determined not to repeat that mistake. The standoff ended when George Pegram, head of Columbia's physics department, intervened, convincing Fermi to back down.[4] Fermi's decision not to share his research would turn out to have far-reaching consequences in the race to beat Hitler to the bomb.

In Germany, Walther Bothe, a future Nobel laureate and friend of Heisenberg, tested graphite for use in the German uranium machine. And he too faced the neutron absorption problem. But he was unable to purify the graphite as Szilard had, so he simply reported that graphite would not work. Heisenberg accepted Bothe's findings and, as he later wrote, "didn't bother to repeat the tests."[5]

Why Bothe failed to purify the graphite is not clear, but his diary holds a possible clue. A year earlier, in June of 1939, he and Heisenberg had sailed together to New York. On board, the forty-seven-year-old Bothe met Ingeborg Moerschner, a woman thirteen years his junior. The two ended up going to the New York World's Fair together and then on to San Francisco where she had a job waiting for her at the German consulate. Bothe returned to Germany but remained smitten. Writing in his diary following the first anniversary of their meeting, he declared that he felt like a drunken teenager. Later, while performing the critical graphite tests, he wrote, "I have been speaking of physics the entire day, while thinking only of you."[6]

With graphite eliminated as an option, Heisenberg decided to use a manufactured chemical called "heavy water" instead. The German Army had occupied Norway several months earlier and seized control of the Norsk Hydro plant, the world's largest manufacturer of heavy water, so Heisenberg seemed assured a plentiful supply. Equipped with the heavy water, the uranium from the Czech mine, and the Belgian Congo uranium which had been stockpiled in German-occupied Belgium, Heisenberg had all the raw material he needed to build what he called a "uranium machine," otherwise known as a nuclear reactor.

Columbia's football team had just started practice for the 1941 season when one of Fermi's colleagues hired several of its players for a special job. With Fermi supervising, they hauled eight tons of uranium, empty canisters, and thirty tons of graphite blocks into the basement of the university's natural sciences building where they assembled them into a large cube.

Uranium ore contained only one U235 nuclear energy storehouse for every 139 U238 atoms. The neutrons that Fermi used to bombard the uranium were far more likely to encounter one of those other atoms or escape through gaps between atoms as than they were to unlock one of the U235 storehouses. To increase the odds of success, Fermi and his colleagues had to take uranium powder (the only form available for purchase), pack it to just the right density and just the right moisture content in canisters of just the right size and shape, and embed those canisters in bricks of graphite of just the right purity, and then position them just the right distance from each other in a graphite structure of just the right shape and volume, then possibly—just possibly—they could slow down enough neutrons and guide them to split enough U235 nuclei to sustain a chain reaction.[7]

Fermi's first attempt to build a reactor in September 1941 fell short of his goal. The generations of split U235 nuclei emitted large numbers of neutrons, but not enough to keep the chain reaction going. Meanwhile, in Germany, Heisenberg felt encouraged by his test results. He would later comment, "It was from September 1941 that we saw an open road ahead of us, leading to the atomic bomb."[8]

CHAPTER THIRTEEN

HEAVY WATER

That same month, British intelligence received word that Norsk Hydro, the electrochemical plant manufacturing heavy water in Nazi-occupied Norway, had shipped forty gallons to a port in Germany. The story of how an obscure industrial plant in the Norwegian wilds came to play a critical role in atom bomb history begins in late 1939, two years before Heisenberg received his first heavy water shipment.

Manufacturing heavy water required enormous amounts of electricity, and Norsk Hydro, built on a granite cliff below a mountain lake, had more than enough. Water from the lake spilled through large steel pipes, turning electricity-producing turbines twenty-four hours a day as it flowed past the plant into a gorge below. Norsk Hydro put that electricity to work, making hydrogen for use in fertilizer. Heavy water was a byproduct which the company sold as a sideline business. Its sales rarely turned a profit, so Norsk Hydro management was surprised when the German chemical company I. G. Farben began placing orders for large quantities.[1] When asked why the sudden interest, the Germans didn't respond. Then, early in 1940, six months after German troops stormed into Poland, Jacques Allier, an official at the Banque de Paris et des Pays-Bas (and an agent of the French intelligence service) showed up at Norsk Hydro with an offer to buy the plant's entire heavy water inventory. Alerted to Germany's purchases, the French physical chemist Frédéric Joliot-Curie had enlisted Allier's help in halting the shipments. But stopping Heisenberg wasn't Joliot-Curie's only motive. He planned to conduct his own nuclear fission experiments, so he too needed heavy water.

Norsk Hydro managers did not know why heavy water was suddenly in demand, but their sympathies lay with the French, so they agreed to sell to Joliot-Curie. Allier planned to ship the heavy water out of Norway as cargo on a regularly scheduled passenger flight. He hadn't expected any problems, but then French intelligence intercepted a message from Germany's spy agency. The Nazis were aware of Allier's mission and had ordered their agents to stop him at any price.[2]

On March 9, a month before the Germans invaded Norway, two trucks departed the Norsk Hydro plant loaded with twenty-six steel flasks full of heavy water. The flasks, specially designed so that they would fit into a suitcase, were delivered to a house in Oslo where Allier was waiting. Using fake names, Allier and a team of French agents reserved seats on two flights scheduled to depart Oslo's Fornebu airport at around the same time. One was bound for Amsterdam; the other, for Perth, Scotland.

On the morning of March 12, Allier and a fellow agent arrived at the airport where they hired baggage handlers to carry their heavy suitcases to the ticket counter. The two men then passed through the boarding gate, crossed the tarmac, and watched as their bags were loaded onto the Amsterdam-bound flight.

The plane took off without incident, but a short while later, two Luftwaffe fighters flew alongside and forced the pilot to land. The Germans had apparently been tracking Allier's movements. On the ground, German troops broke into the plane's baggage compartment and seized the suspect luggage. But instead of heavy water, they found chunks of granite. And when they went looking for Allier, they discovered he wasn't on the flight.

In an impressive act of counter-intelligence legerdemain, Allier and his fellow agents had arranged for their Amsterdam-bound plane to park on the tarmac alongside the plane bound for Perth. The two planes were positioned so that anyone watching from the terminal could not see Allier boarding the Perth flight. Just before the doors shut, a taxi raced up between the two airplanes. One of Allier's agents jumped out, opened the trunk, and loaded suitcases filled with thirteen flasks of heavy water onto the Perth-bound plane. The taxi's driver then sped off with the agent tucked out of sight in the back. The next day, that agent returned and boarded the Perth flight with the remaining thirteen flasks stashed in his suitcase. Two days later, Allier, who had flown to Perth on the earlier flight, showed up in Paris with the full complement of Norsk Hydro heavy water.[3]

The ruse worked, but after German troops occupied Norway in April 1940, the plant's management had no choice but to ramp up production. News of the increased shipments to Germany a year and a half later provided strong evidence that Heisenberg was building a nuclear reactor. If successful, Hitler would be one step closer to acquiring an atomic bomb.

For further evidence of Heisenberg's activities, British intelligence scoured German scientific journals looking for his name. In the past, he'd been a frequent contributor to those journals. Now his name was nowhere to be found. Nor was he listed as teaching a physics course in any 1941 university catalog.[4] His disappearance provided additional evidence that his work was meant to be outside of public view. The findings came as no surprise to Sam Goudsmit. "No one but Professor Heisenberg," his old friend noted, "could be the brains of a German uranium project, and every physicist throughout the world knew that."[5]

While Allied intelligence officials debated how to sabotage the German threat, Norsk Hydro's chief engineer escaped to England, bringing with him microfilm of the plant's blueprints.[6] From then on, cutting the flow of heavy water to Germany became a top Allied priority.

CHAPTER FOURTEEN

LIFE IN LEIPZIG AND BERLIN

The city of Leipzig where Werner and Elisabeth Heisenberg lived with their five children was six hundred miles from the British coast. In 1940 and 1941, it was considered relatively safe from attack by RAF bombers. The food and fuel shortages which would cripple the German economy later in the war were a nuisance, but life for the Heisenbergs was more or less normal. Werner was frequently absent, dividing his weeks equally between Leipzig and the second laboratory in Berlin while Elisabeth stayed home and took care of the children. When together, they still enjoyed special family activities: baptisms, musical evenings with friends, Christmas celebrations, and tending Elisabeth's flower garden. In the summer, Elisabeth, the children, and their nursemaid moved to the family cabin in the Bavarian Alps, and Werner remained in Leipzig or stayed with his in-laws near the Berlin lab while working on the uranium machine.[1]

By 1941, hopes of a quick war lasting only a few months had faded, and oil supplies were starting to dwindle. If Heisenberg succeeded, uranium machines would prove a boon to the German war effort. They could serve as an alternative energy source for German industry and provide power for the nation's fleet of U-boats. A critical part of Hitler's war plan, the stealthy U-boats patrolled the Atlantic, attacking Allied shipments coming from North America, but the time they could spend submerged was limited. Their batteries needed to be charged every day or two and that consumed oxygen, so they had to surface risking detection. Uranium machines wouldn't need oxygen. With engines powered by nuclear reactors,

U-boats would be able to travel long distances without surfacing. Powering U-boats, however, was still secondary to producing plutonium for an atom bomb. Uranium machines could manufacture plutonium far more efficiently than separation plants could enrich U235 for a uranium bomb.

In 1941, Heisenberg had two teams working on building those machines, one at Virus House in Berlin, the other at his university in Leipzig, both wrestling with the same problem: how to maximize the number of split nuclei and freed neutrons. They planned to slow down the freed neutrons by placing their source—several layers of uranium metal sheets—in a tank full of heavy water, but German plants were busy manufacturing metal parts for planes and other war machinery, and there was a shortage of heavy water, so Heisenberg substituted uranium powder for the metal sheets and paraffin wax for the heavy water. Packed with hydrogen and oxygen atoms, the paraffin could block and slow down the speeding neutrons just like the heavy water.

Heisenberg's team at Virus House came up with a cylinder-shaped design that sandwiched thick layers of uranium powder between thin horizontal layers of wax. Meanwhile, the Leipzig team experimented with a spherical prototype. They fit a small metal sphere inside a slightly larger sphere and then fit the two into larger spheres so that they had four spheres nesting inside each other. They then filled the spaces between the spheres with alternating layers of uranium powder and paraffin wax.[2] To test the designs, the two teams immersed their prototypes in a pit of regular water and introduced a neutron source through an opening in the top. They then waited for a chain reaction to begin and monitored the number of freed neutrons produced.

Virus House experimented with five different prototypes. None of them worked. The Leipzig team fared better with their spherical design which yielded Heisenberg's first positive neutron multiplication.[3]

CHAPTER FIFTEEN

THE MANHATTAN PROJECT

"PEARL HARBOR ATTACK," CBS RADIO NEWS, DECEMBER 7, 1941, JOHN CHARLES DALY REPORTING

The Japanese have attacked Pearl Harbor Hawaii by air, President Roosevelt has just announced. The attack also was made on all naval and military activities on the principal island of Oahu.

FRANKLIN D. ROOSEVELT, SPEECH TO JOINT SESSION OF CONGRESS, NEWSREEL, DECEMBER 8, 1941

Yesterday, December 7, 1941, a date which will live in infamy, the United States of America was suddenly and deliberately attacked by naval and air forces of the empire of Japan.

At the time of the Pearl Harbor attack, nuclear fission research in the United States was a leisurely, poorly funded, and loosely coordinated enterprise conducted in university science labs. British scientists, working with refugee scientists from Nazi Germany, Austria, and Hungry, had done much of the early work. That changed with America's entry into the war. Over the next year, the U.S. Army Corps of Engineers organized America's disparate research efforts into a single program. Fission research for the plutonium bomb was consolidated at the University of Chicago in a program code-named "The Metallurgical Lab," and Fermi was notified to get ready to move from New Jersey to Chicago.

Japan had signed a mutual defense pact with Germany and Italy prior to the Pearl Harbor attack. As a result, thousands of individuals of Italian, German, and Japanese ancestry were detained as potentially dangerous "enemy aliens." For Laura and Enrico Fermi, the new classification brought new rules and new restrictions. They couldn't own a shortwave radio, camera, or binoculars or live in a building with other enemy aliens. They were required to obtain permission from the local U.S. attorney before leaving their community.

In April 1942, permission in hand, Enrico boarded the 20th Century Limited train at New York City's Union Station and headed west to his new life in Chicago. Laura and the children would join him at the end of the school year. Officials at the independent agency overseeing wartime scientific research had balked at including Fermi because of his "enemy alien" status and recommended he be investigated more extensively. But Fermi's fellow physicists insisted, and he would soon prove an invaluable member of the team. A brilliant problem solver with an informal, patient manner, he was well liked and an excellent team builder who didn't need to show off how smart he was.[1]

With the children off at summer camp, Laura joined Enrico in Chicago and went apartment hunting. She found a furnished house near the university campus that was available for rent, but the Fermis' "enemy alien" status required their landlord to make some adjustments. He had to remove his shortwave radio and ask the two young Japanese tenants on the third floor to move out.[2]

While Fermi settled in, Glenn Seaborg, one of the U.C. Berkeley scientists who'd discovered plutonium, and his colleague Isadore Perlman boarded the City of San Francisco for the two-day train trip east to Chicago. There they would join Fermi, who was already at work building another reactor. When they arrived, the weather was dark and gloomy, as was their mood.[3]

"The day Iz and I got off the train in Chicago," Seaborg later wrote, "we had little reason to believe America would win. In fact, the evidence pointed the other way.... German engineering was the most respected in the world, especially when it came to arms. Those Panzers rolling across Europe in the blitzkrieg seemed unstoppable.... We were fighting for survival, pure and simple."[4]

The war, once so far away, was suddenly at America's doorstep. In the first six months of 1942, German U-boats prowling waters off the East Coast, the Gulf of Mexico, and the Caribbean sank almost four hundred ships.[5] The United States was not the superpower it would become after the war. People were frightened. And, for the scientists and engineers of the Manhattan Project, the anxiety kept mounting.

In the summer of '42, word spread via the international scientific grapevine that Heisenberg had been named director of the Kaiser Wilhelm Institute for Physics, Germany's leading center for atomic research.[6] That same summer,

Heisenberg gave a lecture at the institute in which he hinted that a uranium burner could be a first step toward the production of a powerful new explosive.[7]

Scientists at the Met Lab in Chicago tried to alert American intelligence authorities, but they didn't seem interested. So they contacted Sam Goudsmit, who had developed relationships with prominent British physicists during the University of Michigan seminars and asked him to alert British intelligence. Goudsmit was told to simply explain to his British friends that Heisenberg's new position "may be of special interest to those working on 'tube alloys.'" "Tube alloys" was code for an atomic bomb.[8]

The leadership position at the Kaiser Wilhelm Institute of Physics had previously been held by Kurt Diebner, the man who had reluctantly drafted Heisenberg to work with the Uranium Club. A loyal Nazi, Diebner had been appointed in 1939 after the previous director, a Nobel laureate, had taken a leave of absence and departed for the United States. KWIP staff didn't consider Diebner of sufficient caliber to lead the institute, so he was named "provisional head."[9]

Not long after Diebner's arrival, several KWIP staffers, allies of Heisenberg, began plotting a palace coup. They suggested Diebner invite Heisenberg as an advisor who would come in from Leipzig one day a week, and the unsuspecting Diebner agreed. Two years later, Heisenberg's plotters persuaded the KWIP governors to dismiss Diebner and name Heisenberg as the director.[10] While in Berlin, he did double duty, also serving as a professor at the University of Berlin.[11]

After his dismissal, Diebner relocated to the army's Gottow research site, a well-equipped explosives laboratory, where he and a cadre of young PhD graduates conducted their own nuclear reactor experiments.[12] His firing aggravated existing tensions between Heisenberg and Diebner. Those tensions would plague Germany's atomic research program for the rest of the war.

Diebner was not a brilliant theorist. Hermann Göring, Hitler's minister without portfolio, described him in a memo as a man "who never got out of the technical college rut."[13] But building a nuclear reactor at that stage didn't require an Einstein. The theoretical work was largely complete. It required a practical scientist skilled in designing and conducting experiments and analyzing data. Heisenberg had never been involved in such a large experimental venture.[14] In fact, he had nearly failed his PhD graduate exam when an experimental physicist questioned him about the resolving power of microscopes and telescopes and asked him to explain how a storage battery worked. That material had been covered in class, yet Heisenberg had trouble answering. Fortunately for Heisenberg, his mentor, a theoretical physics professor, intervened and "vigorously defended his pupil." The future Nobel laureate was allowed to pass with the equivalent of a C, the second-lowest passing grade.[15] His shortcomings as an experimental physicist would not become fully apparent until years later.[16]

In the summer of 1942, Diebner grew tired of waiting for Heisenberg to make progress with his uranium machines, so, without telling Heisenberg, he launched his own experiments. In Heisenberg's cylindrical layered-sandwich design, only the top and bottom of the uranium layer came in contact with the paraffin wax layer. Diebner reasoned if the uranium powder was molded into cubes and immersed in the paraffin wax, all six sides of the cubes would be exposed to the paraffin, dramatically increasing the production of neutrons, the splitting of nuclei, and the release of atomic energy.[17]

To test his theory, he and his crew built a reactor vessel large enough for several men to work in comfortably. Inside they used paraffin wax to construct a wall honeycombed with 6,800 small cube-shaped cells and filled them with powdered uranium. In all, there were 4.4 tons of paraffin and 25 tons of uranium metal powder.

They then lowered the vessel into a pit filled with regular water, inserted measuring devices and a neutron source, and waited . . . and waited for the chain reaction. Diebner's first experiment did not produce a significant increase in freed neutrons, but it worked well enough to convince him that he was on the right track. Meanwhile, Heisenberg doggedly continued with his layered design, varying the number and thickness of the uranium powder layers. None of those experiments improved upon the earlier Leipzig results.[18]

CHAPTER SIXTEEN

A MEETING OF WEAPONS DEVELOPERS

By early June 1942, the surprise German invasion of the Soviet Union had ground to a halt on the outskirts of Moscow. The blitzkrieg war had turned into a war of attrition that the Germans were ill-prepared to fight. Their troops had no gloves, boots, or heavy clothing to get them through the winter, and many had frozen to death.[1] The Germans suffered a second blow in May when the RAF launched a massive thousand-bomber surprise attack on the city of Cologne.

To marshal his resources, the führer ordered a review of all weapons under development. Those that would not be ready for use within nine months would be cancelled.[2] In response, Albert Speer, the architect in charge of War Production, summoned Heisenberg and other weapons developers to Berlin for a meeting.

Standing at a small podium in the front of a large university lecture hall, the first speaker made his case for funding a new type of mine detector. Heisenberg spoke second. The RAF had just begun targeting German cities with those massive raids, so he began his presentation by discussing the new types of bombs made possible by the discovery of nuclear fission. Heisenberg had previously talked publicly about his research on the uranium machine as a possible energy source, but he hadn't mentioned atomic bombs, so his comments created a considerable stir. He explained that there were two possible sources for atomic bombs—U235 and what he called Element 94 (plutonium). He said he was on a path "to explosives that are more than a million times more effective than those currently available" but then switched his focus to the energy-producing uranium machine.[3] He told the assembled officials that his cylindrical reactor had triggered a 13 percent increase in neutrons. Though modest, it was a historic step toward achieving a nuclear chain reaction.[4]

In the question-and-answer session that followed, Heisenberg tried to tamp down the excitement surrounding his mention of a bomb, explaining that building the bomb would require an enormous commitment of resources. With German industry already working at full capacity to meet current demands, an atom bomb would be impossible.

But Speer was sufficiently impressed to offer up to two million Reichsmarks to continue Heisenberg's bomb research. He also volunteered to exempt several hundred soldiers with scientific backgrounds from military service so that they could work on the bomb.[5]

Heisenberg's rival Kurt Diebner was among the attendees at the lecture and was no doubt thrilled. Earlier in the year, he had offered an upbeat assessment of his own research, arguing that the development of an atomic bomb was technically feasible and should be aggressively pursued.[6] But rather than jump at Speer's offer, Heisenberg backed off, explaining that it would take three or four years to build the bomb. By that time, everyone agreed the war would likely be over. Rather than accept the millions on offer, Heisenberg asked for a modest 350,000 Reichsmarks to continue work on his uranium engine, which the navy wanted for its submarines.[7]

Speer approved Heisenberg's funding along with funds to construct an air-raid shelter and laboratory in Berlin for Heisenberg's first full-scale uranium engine.[8] Asked after the war why he rejected Speer's first offer, Heisenberg answered simply, "We wanted to keep things as small as possible."[9] He also claimed—disingenuously—that he played no role in downgrading Germany's atom bomb program, adding that "I was pleased to be spared the responsibility of making a decision."[10]

Oppenheimer, Fermi, and colleagues didn't know it, but Heisenberg's rejection of Speer's offer had effectively killed German chances of building a uranium bomb. Nazi weapons developers could still build a plutonium bomb, but first they needed a working nuclear reactor.

By June 4, 1942, Heisenberg had achieved almost all he had hoped to gain by helping the Nazi regime: draft deferments for him and his staff, guaranteed funding for his research, the regime's recognition of theoretical physics, and an opportunity to pursue exciting nuclear research without risking the führer's wrath should he fail.[11]

During the question-and-answer portion of the Speer meeting, Luftwaffe Field Marshal Erhard Milch asked Heisenberg how large an atomic bomb would be needed to destroy a large city. Heisenberg answered, "about as large as a pineapple."[12] Heisenberg didn't know it, but Milch had a reason for asking. Hitler had been pressuring him to exact revenge against the Allies for the devasting raid on Cologne seven months earlier, which had burned down six hundred acres of the city. Milch and his generals were exploring the possibility of bombing New York City in response. The Germans didn't have a bomber with the range to fly from Germany or German-occupied territory to New York and back, so they were exploring the possibility of launching a heavy bomber from an aircraft carrier with a light bomber secured underneath. Once in range of New York, the light bomber would detach from the mother ship, fly the rest of the way, drop its bomb, then ditch in the Atlantic where a submarine would pick up the crew. A pineapple-sized atomic bomb would be the perfect weapon for such a fantastic scheme. But the idea never advanced beyond the talking stage.[13]

On June 23, 1942, Speer met with Adolf Hitler to review the results of the weapons development conference. Nuclear fission was item sixteen on a long list of issues and received only a brief mention. According to Speer, it was the only time in all his meetings with the führer during the war that he ever discussed the matter. "The idea [of an atom bomb] quite obviously strained his intellectual capacity," Speer noted in his memoirs.[14]

While Speer was meeting with Hitler, Heisenberg's colleague Robert Döpel was in the Leipzig lab conducting an experiment with *L-IV*, their fourth Leipzig spherical uranium machine prototype. It didn't go well. After twenty days immersed in a water tank, *L-IV* suddenly began emitting bubbles. Analysis showed that the bubbles contained hydrogen, which was created by a chemical reaction between the uranium metal powder and water that had leaked into the machine. Döpel ordered his crew to hoist the machine out of the tank so they could locate the leak. A technician unscrewed one of the inlet caps, and suddenly a stream of hot gas containing particles of burning uranium powder shot out, followed by a foot-long flame that melted the sphere's aluminum casing and set more uranium powder on fire. Döpel poured water on the fire, trying to put it out, but it continued for a few terrifying seconds before it finally subsided.

A panicked Döpel called Heisenberg, who was nearby tutoring one of his students. When he arrived, they pumped out the precious heavy water and winched the sphere back into the tank to cool down. Confident that the crisis had been averted, Heisenberg left to go back to his student.

But the crisis had just begun. During the next hour, the sphere's temperature continued to rise, and Heisenberg was called again. He and Döpel were peering down into the boiling water when the sphere suddenly started to shudder and swell. They ran for the door, and seconds later, an explosion ripped through the laboratory, spewing burning uranium into the lab and throughout the building.[15]

CHAPTER SEVENTEEN

GENERAL LESLIE GROVES

The United States had no shelter, armor, or fortification strong enough to protect the American people from an atom bomb. The nation's best option was to counter the German threat with a bomb of its own. To spearhead that effort, the government turned to the U.S. Army Corps of Engineers and General Leslie Groves who set up a new organization code-named the Manhattan Engineering District, commonly known as the Manhattan Project. It would be overseen by a four-member Military Policy Committee (MPC) with Groves as its executive officer.[1] In practice, the MPC would end up rubber-stamping virtually everything Groves recommended.[2]

A forty-five-year-old army engineer, Groves was an abrasive, egotistical, no-nonsense manager with no patience for subtlety or diplomacy. He was "the biggest son-of-a-bitch I ever met in my life," said one Groves subordinate, "but also one of the most capable individuals."[3]

The son of an army chaplain, Groves graduated fourth in his class at West Point and had several years' experience tackling large construction projects. He was just finalizing work on the Pentagon, at the time the world's largest building, and was looking forward to an overseas assignment when his superior officer brought him some unwelcomed news.

"The secretary of war has selected you for a very important assignment, and the President has approved the selection," he told Groves, adding, "if you do this job right, it will win the war."[4] No stranger to the workings of government bureaucracy, Groves didn't need to be told who would catch the blame if things went wrong.

Not pleased, Groves worried that he would spend the rest of the war wrangling a bunch of university professors in a project whose chances of success were uncertain at best. More importantly, he feared the assignment would deprive him of battlefield experience, the sort of experience the army valued when handing out promotions. "I was probably the angriest officer in the United States Army," he later commented.[5]

Groves knew the United States was attempting to build an atom bomb. He had been asked by the project's previous head to recommend companies with a track record and the expertise to handle the tasks that lay ahead, but he knew little of the details.[6]

Groves's superior assured him that the project was already well advanced and that the R&D was complete.[7] After nearly a quarter century in the army, he knew to be skeptical. Groves accepted the assignment with one request, that he be promoted to brigadier general. In his autobiography, *Now It Can Be Told*, Groves explained, "I thought that my position would be stronger if [the scientists] thought of me from the first as a general instead of as a promoted colonel."[8] Groves was extremely ambitious. The assignment offered an opportunity to move up the ranks, and he was not going to pass it up. As a result, he now outranked several field officers who had once been his superiors. One had been a brigadier general when Groves was a West Point cadet. Another had served on a board that promoted second lieutenant Groves to first lieutenant.[9] Needless to say, they were not thrilled.

The freshly minted General Groves wasted no time getting started. He and his secretary Jean O'Leary, soon to be promoted to chief administrative assistant, commandeered two rooms in the new War Department building. They sealed the ventilation louvers on the doors so their conversations couldn't be heard in the hall and brought in some oak desks, metal file cabinets, and a safe for top-secret papers and got down to business.[10] (Never a fan of large unwieldy bureaucracies, Groves did his best to keep the operation small, but was eventually forced to annex four more rooms.)

In just three days, General Groves finalized a directive left unsigned by his predecessor to buy ninety-two square miles of semi-wilderness in Tennessee where he would build uranium separation plants. He also marched into the offices of the War Production Board, the agency that decided who got first claim on construction materials in short supply due to the war. He handed its civilian head, Donald Nelson, a letter he had drafted granting the Manhattan Engineering

District the top AAA rating, and, when Nelson objected, Groves threatened to call President Roosevelt directly.

Nelson signed the letter. Winding up the whirlwind first three days, Groves arranged to purchase 1,200 tons of uranium ore stored in a Staten Island, New York, warehouse. It had sat there for more than a year while its owner searched unsuccessfully for someone to buy it. Edgar Sengier, the chairman of a Belgian multinational mining company and the owner of the uranium cache, had quite a story to tell.

Uranium was used as a source of radium for the luminous paint that made watch dials, clocks, and airplane instruments glow in the dark. Three years earlier, the demand for radium had plummeted—and consequently the price of uranium, so Sengier had shuttered most of the company's mines. But in May 1939, on a visit to London, Sengier met a distinguished British scientist who surprised him with a request that he grant the British government an option to buy all his uranium. A shrewd businessman, Sengier refused.

A short while later, he traveled to Paris where he met the Nobel laureate Frédéric Joliot-Curie, whose work inducing artificial radioactivity had inspired Fermi's Nobel Prize–winning experiments. Joliot-Curie informed him that German scientists had split the uranium nucleus and discovered atomic energy. He then shocked Sengier by proposing that the two men partner in a venture to build an atom bomb. He suggested they set up a laboratory in the Sahara Desert where they could develop the bomb ahead of the Germans, British, and Americans.[11] As a young man, Sengier had spent five years in mountainous China building tramways. The scheme appealed to the adventurous streak of his youth, and he agreed.

But then World War II broke out, and everything changed. Sengier fled his native Belgium, taking up residence in the United States. Fearing a German invasion of the Congo, he arranged to have 1,200 tons of uranium ore loaded on two freighters and shipped to New York. There it sat while Sengier tried unsuccessfully to interest the State Department in buying it. General Groves learned about the orphaned uranium his first day on the job and immediately ordered an assistant to go to New York and buy the entire stock.[12]

The Manhattan Project was a top-secret operation with a virtually unlimited budget whose workforce would eventually total more than half a million people: scientists, engineers, military personnel, government bureaucrats, union bosses, and civilian workers. Keeping them all in line would require some dexterous juggling.

Groves's relationship with the scientists was never close. He would later describe the Manhattan Project as a "two-billion-dollar grand opera" (twenty-six billion in today's dollars), in which he was the "impresario" and they, "thousands of temperamental stars from all walks of life."[13]

In early October 1942, after touring his newly purchased land in Tennessee, the impresario began a cross-country trip to meet some of those "temperamental stars." He needed them to engineer a solution to what, at the time, was the project's most daunting challenge: separating U235 atoms from U238 atoms.[14] Six years earlier, Fermi had discovered a way to breach nature's defenses and gain access to U235's nuclear energy. Now, mankind was knocking at the gates looking for a way to harvest that energy. And once again nature was resisting. Scientists had a proven method for separating *different* elements which had different chemical structures such as gold and silver, but it wouldn't work with U235 and U238 because the two were both uranium atoms. The only difference: U235 had three fewer neutrons than U238 and about a 1 percent smaller mass. Fermi had used a chink in nature's armor—U235's vulnerability to slow-moving neutrons—to split the uranium nucleus. Manhattan Project engineers would now attempt to exploit an infinitesimal difference in mass to separate the two atoms.

Imagine a vast desert stretching as far as the eye can see made up of trillions and trillions and trillions of silky fine grains of sand. Imagine that those grains of sand are atoms. To build an atom bomb, Manhattan Project engineers had to develop a technology to sort through all those trillions of atoms, identify the one U235 atom out of every U238 atom that was infinitesimally smaller than the others, and then figure out a way to extract them. Harvesting enough U235 to fuel a bomb would require an extraordinary amount of processing: Twenty-four hours a day for weeks? Months? Years? Nobody knew.

Groves had been given the names of five scientists who might be up to the challenge. After the first two proved to be disappointments, he headed to Columbia University in New York to meet candidate number three.[15] Dressed in civilian clothes so as not to attract attention, he and two of his staff met with Dr. Harold Urey, the Nobel Prize–winning head of Columbia's nuclear laboratory, and Dr. John Dunning, the man who supervised the lab's day-to-day operations.

At Columbia, scientists were exploring a separation technology called gaseous diffusion, which had been developed in England based on a theory called

Graham's Law. They would convert uranium to a gas and pump it through miles and miles and miles of pipes connected to what they called "diffusers," which were essentially porous tubes inside airtight vacuum chambers. The vacuum would suction the lighter U235 atoms out of the porous tube and into the chamber at a faster rate than the larger U238 atoms, resulting in an infinitesimal increase in the concentration of U235. The gas inside the chamber would then be pumped into the next chamber and then into some hundreds of additional chambers until the concentration of U235 atoms gradually increased to weapons grade.

In the beginning, the concentration of U235 atoms would be less than 1 percent. Weapons-grade uranium would have to be 80 percent or more. In September 1942, no one knew how much enriched U235 would be required—ten, one hundred, one thousand pounds?—but they feared the Germans were far ahead, so they had no choice but to proceed. To speed up the process, General Groves and the Columbia engineers decided to build a massive plant with some hundred miles of pipe and three thousand vacuum chambers.[16] Groves, who had just completed construction on the six-million-square-foot Pentagon in Washington, D.C., had built many large structures in his career, but the challenges involved in constructing the proposed gaseous diffusion plant were unlike any he had ever faced before.

The entire complex of pipes, pumps, and vacuum chambers would have to be airtight. A single pin-size hole or defective weld could allow moisture from the outside air to seep in, causing the uranium gas to decompose and turn into a powder, clogging the billions of microscopic holes in the tubes that were essential to the separation process.[17] In addition, uranium gas was highly corrosive. Pumped seven days a week for weeks and months on end, it would eat away at the metal pipes. And then there was the problem of lubricating the seals on the pumps. Grease and uranium gas didn't mix. When they came in contact, they produced a violent reaction.[18]

Solving those problems would prove difficult, but nowhere near as challenging as designing and manufacturing the porous tubes that strained out the U238 atoms. The tubes would be made of thin metal sheets with billions of microscopic holes per square inch. The size of the holes could not exceed one ten-thousandth of a millimeter. The metal had to be strong enough to resist the constant pressure of the gas flow and flexible enough so that it could be rolled into a tube.[19]

American industry had nothing on the shelf that could come close to meeting those requirements.[20] In his autobiography, Groves commented, "The problems would have taxed the ingenuity and industry of Hercules."[21] When he left New York, the general doubted that gaseous diffusion would ever work.

His next stop was the University of Chicago and the Metallurgical Lab (a name chosen to disguise its secret mission) where Fermi and Seaborg were working on plutonium. The Met Lab scientists had been debating for months the best method for cooling the giant nuclear reactors that would manufacture the plutonium.[22] General Groves gave them one week to make up their minds. He then asked how much plutonium would be needed for a critical mass to trigger the explosion. Their estimate seemed plausible, and they all agreed, so Groves accepted it, adding one more question: how accurate was the estimate? "I expected a reply of 'within twenty-five or fifty per cent' . . . but I was horrified when they quite blandly replied that they thought it was correct within a factor of ten."[23]

A factor of ten meant an estimate of one hundred pounds of plutonium could end up being as little as ten pounds or as much as a thousand pounds. Without a more accurate number, Groves had no idea how many nuclear reactors to build. "While I had known that we were proceeding in the dark, this conversation brought it home to me with the impact of a pile driver," Groves later acknowledged.[24]

Before leaving Chicago, General Groves asked Arthur Compton, the head of the Met Lab, how powerful the atom bomb would be. The British had a ten-ton bomb in its arsenal—the "Grand Slam"—with an explosive force equal to five hundred tons of dynamite. Groves expected something similar.

Compton estimated somewhere between five and ten thousand tons of dynamite. Groves was floored. The atom bomb was not just an enhanced version of existing ordnance. It was a revolutionary leap into a realm of war technology never seen before. Groves had been in the army long enough to know that no military commander would ever be given sole authority to use such a powerful weapon. It would have to be a political decision, and he worried that mixing politics and war could only spell trouble.[25]

As he boarded a train for San Francisco, his next stop, General Groves felt blindsided and exasperated. It seemed that the work on U235 separation technology at Columbia and on plutonium manufacturing in Chicago had barely begun. He had been sold a bill of goods. The research and development was nowhere near completed.

His spirits were buoyed, however, by reports of a promising electromagnetic method for separating U235. Its inventor, U.C. Berkeley professor Ernest O.

Lawrence, was waiting for Groves when he got off the train. Lawrence impressed Groves immediately. Energetic and confident, he promised Groves, "unlike all those theories in Chicago and at Columbia, . . . here you'll see separation actually going on."[26]

A scientist and engineer, Lawrence was known worldwide as the inventor of the atom-smashing cyclotron. After America's entry into the war, Lawrence realized that his invention, previously used to explore atomic nuclei and to diagnose and treat cancer, could be modified to separate U235 atoms from U238.[27]

Eager to see Lawrence's invention, Groves insisted they drive directly to the U.C. Berkeley campus where Lawrence led him into a domed building housing a prototype of his Calutron (a combination of California, university, and cyclotron). Inside, a giant magnet rose twenty feet above the floor. Nestled in the middle between the positive and negative poles was a vacuum chamber where the separation took place. The calutron fired a beam of electrically charged uranium atoms into the chamber. Once inside, the magnet bent the beam upwards so that its path changed from a straight line to an arc. The U238 atoms had a larger mass than the U235 atoms, so their arc was slightly greater than that of the U235. The separation was very small, only about a quarter of an inch, but the two paths diverged just enough so that the atoms could be collected in separate pockets at the end of the path.[28]

Groves was relieved to see a prototype that actually worked. He asked Lawrence how much U235 he had separated. "Well, actually, we don't get any sizeable separation at all," Lawrence answered. "This is still all experimental, you see."[29]

Groves was horrified. In his autobiography he commented, "It seemed as if the whole endeavor was founded on possibilities rather than probabilities. . . . Never in history has anyone embarking on an important undertaking had so little certainty about how to proceed as we had then."[30]

Prior to taking on his new assignment, Groves had worked with a long list of blue-chip companies including Du Pont, Tennessee Eastman, Allis-Chalmers, Union Carbide, and General Electric, but finding ones that could invent the new technologies, build, and run the massive industrial plants necessary to fuel an atom bomb, and do it in time to beat the Germans, was without a doubt his greatest challenge. In peacetime, according to one estimate, it would take ten to fifteen years to build Lawrence's electromagnetic separation plant.[31] With the country at war and Heisenberg enjoying a two-year head start, Groves did what would become common for him, he set an impossible deadline: summer 1945—just under three years—to process the uranium and build the first bomb.

On October 30, 1942, Groves called a senior Du Pont executive with whom he'd worked in the past and asked him to go to Chicago to evaluate the work Fermi and the others were doing with plutonium. A week later, Groves phoned him for an update. The executive did not mince words: "The whole thing seems beyond human capability. I don't think there's a Chinaman's chance of getting this thing done" in the time allotted.[32]

Undeterred, Groves arranged a meeting with Du Pont's president, who arrived with a list of reasons why Du Pont should not get involved: The company was overcommitted. The work was dangerous. Du Pont had no experience building reactors. And there was not enough time.

Groves countered with an appeal to patriotism: "If we get the weapon first, then we'll have a way of shortening the war and saving tens of thousands of lives." The hard-charging general would spend the next three years pushing, prodding, cajoling, and bullying CEOs, government officials, and pretty much anyone else who got in his way, turning what had been a leisurely paced academic exercise into a crash military-run campaign. Groves didn't care if he alienated people. No one, including his staff whom he screamed at and demeaned in public, escaped his bullying. In this case, the strong-arm tactics weren't necessary. The appeal worked. "We are an American company," the executive told him. "Du Pont's assistance is needed—we cannot refuse to accept the work."[33]

In December 1942, with Du Pont on board, General Groves traveled to New York City for a dinner with Dobie Keith, the forty-two-year-old vice president of M. W. Kellogg, a large engineering company. On the menu: finalizing plans for construction of the mammoth U235 gaseous diffusion separation plant. Keith was considered one of the country's top chemical engineers.[34] A manager with imagination and an impressive history of transforming theory into reality, he struck General Groves as the perfect man for the job.

At dinner, Groves was his usual gruff self: no small talk, no smiling, no effort to please. But as they left the restaurant and walked together, "He changed into a human," wrote veteran journalist and atom bomb historian Stephane Groueff. He talked about himself, his service in the engineer corps, his student years at West Point, and his anger over being denied overseas assignment. And he shared the argument that ultimately changed his mind. "If you're successful," Groves's commanding officer told him, "it will shorten the war."[35]

When Keith accepted Groves's invitation for dinner, he had a good idea why the general had called. He had done preliminary work on gaseous diffusion with Columbia University. He knew the odds against success were astronomical and was inclined to refuse the assignment. But, whether it was sincere or a ploy to bring Keith on board, General Groves's attempts at getting personal worked.[36]

"I can't work unless I have complete responsibility and no interference from anyone," Keith told the general. Surprisingly, Groves agreed. "You're the boss. I promise I won't interfere, and neither will anyone else." The two shook hands, and in early 1943, Keith set up Kellex, a company independent from M. W. Kellogg, and began assembling an all-star team of engineering experts to tackle the daunting list of unsolved and unprecedented problems.[37] The fledgling Kellex corporation would soon grow to employ three thousand people.

CHAPTER EIGHTEEN

GENERAL GROVES MEETS ROBERT OPPENHEIMER

General Groves met the future scientific director of the Manhattan Project's Los Alamos Lab, Julius Robert Oppenheimer, at U.C. Berkeley on his initial inspection tour of project research facilities. A gaunt, chain-smoking intellectual, Oppenheimer was the son of a wealthy, German-born Jewish textile importer.

Robert Oppenheimer and General Leslie Groves. DIGITAL PHOTO ARCHIVE, DEPARTMENT OF ENERGY, COURTESY AIP EMILIO SEGRÈ VISUAL ARCHIVES.

Groves liked what he saw in the thirty-eight-year-old physics professor. "He's a real genius," he told an interviewer. "He can talk to you about anything you bring up . . . except sports."[1] Groves had other business to attend to, so he asked Oppenheimer to come to Chicago and join him during the train ride back to Washington. Hours of discussion in Groves's private compartment sealed the deal.

Oppenheimer had worked at U.C. Berkeley's radiation lab where information was shared on a need-to-know basis. He found the compartmentalization stultifying and recommended to Groves that Manhattan Project scientists all work in one isolated area where they could share ideas freely without worrying about security, an important contrast to Heisenberg and colleagues who insisted on working at their different universities. Oppenheimer even recommended a location—the New Mexico desert where he had spent his summers and where he owned a 154-acre ranch.

Oppenheimer was brilliant, but he'd never managed anything larger than a university seminar. His medical history included serious bouts of depression including a near suicide, and his leftist politics raised concerns that he might be a security risk and pass atomic secrets to the Soviets. Oppenheimer admitted to donating to many Communist-front organizations. He'd never joined the party, but his wife Kitty, brother, sister-in-law, and former fiancée had all been members. Kitty had, in fact, been married to one of the Communist Party's American leaders.[2]

Even before his name surfaced in connection with the Manhattan Project, he had come to the attention of the FBI, which was investigating suspected Soviet espionage at U.C. Berkeley's radiation lab. When he became a candidate for the Manhattan Project scientific director position, the FBI passed his file to Colonel Boris Pash, the San Francisco area chief for army counterintelligence.[3] The son of a Russian émigré, Pash had fought against the Bolsheviks in the Russian Revolution. He was a zealous, some said fanatical, anti-Communist. And he vehemently opposed Oppenheimer's appointment.

Oppenheimer's political past troubled General Groves as well. General Groves was as fanatical about security as Pash was about Communists, but he prided himself on his ability to judge people. He knew that Oppenheimer and Kitty were ambitious, that Oppenheimer yearned for a Nobel Prize, but would probably never get one. Physicists generally did their best work in their twenties. At thirty-eight, Oppenheimer was over the hill.[4] If he succeeded, the atom bomb would bring him something even more prestigious: a place in world history.

Groves trusted that would be sufficient to ensure Oppenheimer's loyalty. During Groves's tenure as head of the Manhattan Project, circumstances compelled him to place bets, making numerous decisions based on little evidence. None more consequential than taking a chance on Oppenheimer.

Groves made the appointment over Pash's objection, but Pash never let up. He kept Oppenheimer under constant surveillance. He tapped his phone, opened his mail, assigned counter-intelligence agents as his driver and bodyguard, and repeatedly interrogated him. Oppenheimer cooperated with every request, but no matter what he did, he could not convince Pash of his loyalty.

With Oppenheimer in place, the Allied race to build an atom bomb was launched. Oppenheimer would lead the Allies; Heisenberg, the Germans. Despite being on opposite sides, the two had much in common. They were both university professors, theoretical physicists, and about the same age. Both came from privileged, educated families. (Heisenberg's father was a professor of medieval and modern Greek studies.) Both had been investigated by the state as subversives: Oppenheimer, for supporting Communists; Heisenberg, for defending "Jewish science." And both owed their exoneration to friends in high places: Oppenheimer, to General Groves; Heisenberg, to SS Commander Heinrich Himmler.

CHAPTER NINETEEN

OPERATION FRESHMAN

Oppenheimer and Heisenberg had one more thing in common: a keen interest in the output of the Norsk Hydro plant in Nazi-occupied Norway. By the beginning of 1942, the company had increased its production of heavy water tenfold to three hundred pounds a month and was making plans to boost output even further. Word of the stepped-up production convinced British intelligence that it couldn't wait.[1] It had to prevent that heavy water from reaching Heisenberg. British Special Operations Executive (SOE) rejected an RAF air raid, which would likely cause numerous civilian casualties, and opted instead for a daring attack by an airborne team of saboteurs. The plan, code-named Operation Freshman, would use two aircrafts, each towing a glider to fly a team of saboteurs under the cover of darkness to a remote landing strip prepared by Norwegian resistance fighters. From there, they would ski about fifteen miles, sneak into the plant, set their explosive charges, and then hike two hundred miles to the Swedish border.[2]

On November 19, 1942, two four-engine Halifax bombers towing the fragile wooden gliders took off from Wick airfield in Scotland bound for Norway. Each glider carried fifteen commandos, two crew members, and several hundred pounds of explosives and equipment. To familiarize themselves with the terrain, the bomber and glider crews had conducted a test run the night before, flying four hundred miles across the North Sea and searching for a seven-hundred-foot landing zone marked by the resistance fighters. It didn't go well. The weather was clear, but they couldn't find the site.[3]

The next day, the day of the mission, the problems began almost immediately. Just after takeoff, both bombers lost communication with their gliders. The first bomber wasted valuable fuel searching for the landing zone in the snow-covered landscape and was forced to abort the mission. After turning around, it ran into icy conditions near the Norwegian coast, causing the glider's tow rope to snap. The bomber pilot radioed base that the glider had crashed into the sea.

The second bomber, meanwhile, came in low over the North Sea to avoid the clouds, planning to climb once it reached clear skies over Norway. But ten miles inland, the bomber crashed into a mountain. The four-man bomber crew and three of the commandos in the glider were killed. Several other commandos were severely injured. When German troops located the wreckage early in the morning, they found fourteen survivors and, in the wreckage of the glider, ski equipment and tents, machine and tommy guns, radios, food, and explosives.

After taking the survivors to headquarters, the Germans interrogated them just long enough to get their names, rank, and service number. That evening, they shot and killed all fourteen. It was obvious from the explosives on board that they were on a sabotage mission, but the exact target wasn't known until the next day, when the Germans found the second glider. It hadn't crashed in the sea as first thought. It crashed on land.

The officers who interrogated the first survivors had been reprimanded for their hasty execution of the saboteurs.[4] This time they grilled them and tortured several at length, forcing them to divulge their target. Then they killed them.

The raid had been a costly failure: thirty-eight lives were lost, and the plant was still up and running.[5] And, if they didn't know it before, the Nazis were certainly aware now that the Allies had made disabling the Norsk Hydro facility and halting Germany's atom bomb development a top priority.[6]

In November 1942, as the Operation Freshman disaster was unfolding, Werner Heisenberg was waiting for the delivery of metal uranium plates for his most ambitious reactor experiment to date. *B-VI*, his sixth Berlin experiment, would be a full-scale prototype built in an underground bunker now under construction. German industry was working day and night to meet the demands of the military services while minister Speer supervised the prioritizing of those demands. After Heisenberg turned down Speer's offer of two million Reichsmarks for developing the uranium machine, Speer assigned Heisenberg's project a low-priority rating. Heisenberg appealed for an upgrade to speed up delivery, but Speer refused.[7]

While waiting for materials, Heisenberg kept busy pursuing his duties as university professor in Berlin. He taught theoretical physics twice a week, worked with postgraduates on their research, and oversaw a lecture series on cosmic radiation. He also found time to assist the Reich with its propaganda efforts outside the country. After the German Army overran most of Europe, the occupied countries were forced to open German cultural institutes, which promoted visits and lectures by distinguished German academics. Those goodwill ambassadors became an important part of Germany's external propaganda program designed to impress audiences with the quality of German culture and encourage support for the Nazi regime.

Heisenberg, like many, if not most, Germans, was not enthusiastic about the war, but he was a patriot, and when approached to become part of the program, he agreed under one condition. No fan of National Socialism, he insisted that his lectures be about science, not politics.[8] He traveled in 1942 to neutral Switzerland to lecture and to Budapest when its loyalty to the Reich had begun to weaken.[9] In October 1943, he was sent to the German-occupied Netherlands following a summer of protests by Dutch students and professors over German occupation policies and the deportation of Jews to death camps.[10] He avoided politics as much as possible when lecturing, but during a walk with an old colleague and friend, Hendrick Casimir, he let loose about his views on history, world politics, and the abhorrent conduct of his fellow countrymen. "Only a nation that rules ruthlessly can maintain itself," he said. "Democracy cannot develop sufficient energy to rule Europe. There are, therefore, only two possibilities: Germany and Russia, and perhaps a Europe under German leadership is the lesser evil."[11] Casimir would repeat that conversation after the war, and it would become a key piece of evidence for scholars trying to understand Heisenberg's mindset during the war.

CHICAGO

Fermi planned to test his latest prototype reactor in a forest twenty miles southwest of Chicago. He was confident that a runaway chain reaction spewing deadly radiation was an impossibility, but to be safe he opted for the extra measure of protection. Unfortunately, a labor strike delayed work on the reactor building. Worried about losing more ground to Heisenberg, Fermi decided not to wait. He suggested an alternate location: a squash court inside the abandoned University of Chicago football stadium. Fermi and his Manhattan Project bosses knew the university president was not likely to approve, so they decided not to inform him of their plans.[1]

December 2, 1942, the day of the test, was bitterly cold. Outside the stadium, traffic was lighter than usual due to the recent introduction of gasoline rationing. Inside, beneath the stadium's west stands, Fermi, bundled in his coat and hat, picked up where he left off in New York. But this time, the stakes were much, much higher.

The United States was at war, and American soldiers were dying in battle. The U.S. State Department had just confirmed rumors of a Nazi extermination campaign against Jews in Nazi-occupied Europe.[2] An estimated two million men, women, and children had been killed, and Laura Fermi was desperately worried about the fate of her brother, sisters, and father, whom she and Enrico had left behind in Rome. For the former University of Rome physics professor, building a working nuclear reactor was no longer an academic exercise; it was a critical first step in manufacturing a plutonium bomb and winning the war.

Fermi's Met Lab associates and some thirty high school dropouts waiting for their draft notices had spent almost three weeks hauling tons of graphite bars into the stadium so that skilled craftsmen could cut them to proper length and drill every fourth bar with holes to hold the uranium.[3] In all, Fermi's crew cut forty-five thousand bars with nineteen thousand holes. It was grimy, grueling work. Graphite dust filled the air. It dirtied faces and hands, soiled lab coats, and left a slippery coating on walls and floors.[4]

When it was finished, technicians made their final preparations while Fermi and his half-frozen colleagues watched from a balcony overlooking the unheated squash court. Below, George Weil, a young Met Lab scientist, waited for Fermi's signal to remove the cadmium rods which were soaking up neutrons preventing a chain reaction. Fermi had built thirty prototypes to prepare for this day. He was confident this one would work. Not quite so certain, onlookers watched in tense silence. Among them was Fermi's Columbia University colleague Leo Szilard.

Several months earlier, Szilard had received a vaguely worded telegram from a scientist who knew Heisenberg. It was very likely Fritz Houtermans. The message, sent from neutral Switzerland, has been lost, but Szilard briefed the director of the Chicago Met Lab on its contents, and the director wrote a memo stating: "the Germans have succeeded in making the chain reaction work. Our rough guess is that they may have had the reaction working for two or three months." The memo warned, "there is a real danger of bombardment by the Germans within the next few months using bombs designed to spread radioactive material in lethal quantities."[5]

Months later, Houtermans would send a cryptic telegram to a friend at the Met Lab in Chicago. Apparently worried that Nazi agents might intercept it, he wrote simply, "Hurry up! We are nearly there."[6]

Fermi and colleagues had presumably been briefed about the first telegram and the memo. An atom bomb in the Nazi arsenal, even a dirty bomb that killed by spreading radioactive material, could alter the balance of military power in Europe and throughout the world for generations. The specter of such a weapon poisoning London, New York, and other Allied cities would haunt Allied scientists all during the war.

The American embassy in London had warned British intelligence of the threat. But the Brits didn't take the matter seriously. They claimed that Szilard was wrong, that Heisenberg had not built a working reactor, and that Heisenberg's research was still in progress.[7] But the British didn't reveal what made them so sure. When it came to nuclear fission research and scientific intelligence gathering, the two allies didn't quite trust each other.[8] Unable to verify the British intelligence but knowing Heisenberg and his fervent devotion to the fatherland, Fermi and Oppenheimer assumed the worst.

On Fermi's signal, George Weil removed all but one rod from the reactor. After checking his calculations, Fermi signaled Weil to begin inching out that last one. Inside the reactor, split uranium nuclei spewed out thousands, then millions, of neutrons. A device similar to a Geiger counter emitted an arrhythmic clicking noise that soon transformed into a steady static. As the spectators held their breath, Fermi remained calm, stopping every few minutes to compare the readings to his estimates. He had spent months preparing for this moment. And, according to Fermi's former student Herbert Anderson, he was confident he would succeed. "He was very much the showman on that day . . . He had a lot of people watching . . . He just enjoyed it tremendously."[9]

Over the next hour, the clicking intensified as the shivering spectators stamped their feet to keep warm. When the intensity grew too great for the neutron counter, a chart recorder took over, plotting the neutron activity on a rotating cylinder of graph paper. Each time Weil pulled out the rod, the pen drew a line that rose and then leveled off, and Fermi consulted his notes to make sure the levels were as predicted. Then, just as the monotony of the repeated measuring had lulled the onlookers into a state of somnambulant boredom, a loud crack coming from the reactor caused everyone to jump. Something was wrong. An automatic safety device had been tripped, plunging a control rod into the heart of the reactor and killing the chain reaction.

As the startled balcony crowd gasped and prepared to flee, Fermi calmly consulted his calculations. He knew exactly what had happened. The neutron counter on the safety device had been set too low. Neutron activity inside the reactor had exceeded its setting. Unflappable as ever, Fermi checked his watch—it was 11:30 a.m.—and called for a lunch break.[10]

At 2:00 p.m., Fermi picked up where he left off. The last control rod was set at the point where it had triggered the shutdown. Fermi checked his calculations once again and motioned for Weil to pull the control rod out another twelve inches. Again, the neutron intensity increased. But this time it kept increasing. "This is going to do it," Fermi announced.[11]

The clicking of the neutron counter grew louder and faster, and the line on the chart recorder kept rising, then leveled off. The reactor had gone critical. One or more neutrons from each split U235 nucleus had managed to find and split another U235 nucleus. The chain reaction was self-perpetuating.

At 3:49 on the afternoon of December 2, 1942, the clicking of Fermi's neutron counter signaled a nuclear future that Einstein and H. G. Wells had only imagined, a future that was at once exhilarating and terrifying. Arthur Compton, the Met Lab director, phoned the other members of the Military Policy Committee with a message: "The Italian navigator has just landed in the new world."

"Were the natives friendly?" the listener responded.

"Everyone landed safely."[12]

Fermi let the reaction run for four more minutes before reinserting the rods and shutting it down. The relieved spectators uncorked a bottle of Chianti and signed the bottle's straw covering. Those barely decipherable signatures are the only record of the forty-two people who witnessed that historic moment. Fermi and his colleagues had been preparing for that moment for a long time, but according to Eugene Wigner, a Hungarian-American physicist working on the Manhattan Project, they still had an "eerie feeling when we knew we had actually done it."

Leo Szilard was frightened. "I shook hands with Fermi, and I said I thought this day would go down as a black day in the history of mankind."[13]

The reactor, a massive structure twenty feet tall, twenty-five feet wide containing 771,000 pounds of graphite and 93,000 pounds of uranium, generated only half a watt of energy, barely enough to light a flashlight bulb, but enough to keep the Allies within striking distance of Heisenberg.[14] The United States, Szilard estimated, was now only two or three months behind the Nazis.[15]

CHAPTER TWENTY-ONE

SS *MONT-BLANC*

No nation had ever wielded a weapon as powerful as an atomic bomb. How much energy would it generate? How wide a swath of destruction would it leave? One mile? Two miles? How many lives would it take? Again, more questions in search of answers. The science of nuclear fission was still in its infancy. There were no precedents. "Most experience in life can be comprehended by prior experiences," the physicist Norris Bradbury would later write, "but the atom bomb did not fit any preconceptions possessed by anybody."[1] The largest explosion ever to hit a populated area came from an accidental blast in Canada during World War I.

On December 6, 1917, a French munitions ship, the SS *Mont-Blanc*, sailed into the harbor in Nova Scotia. From there, it planned to join a convoy of ships sailing to a port in France. On board were 2,300 tons of picric acid, a powerful explosive; 200 tons of TNT; and numerous drums of high-octane fuel.[2]

At 9:04 that morning, while negotiating a narrow passage into the harbor, the *Mont-Blanc* collided with a Norwegian ship setting off a powerful blast that destroyed two square miles of the north end of the city, ignited dozens of fires, and sent missiles flying three to four miles. The explosion killed more than 1,500 people instantly and injured thousands more.[3]

Hospitals filled to overflowing. Freight cars two miles away were blown off their tracks. Telephone and telegraph facilities were downed for thirty miles around the city. The *Toronto Daily Star* reported: "Screams of suffering children stabbed the heart." It appeared "as if some giant scythe had blown down [the] neighborhood and then spat fire on it."[4]

It took more than two thousand tons of high-energy chemical explosives to level those two square miles of Halifax waterfront. How would an atom bomb compare? Oppenheimer used that explosion to estimate the damage from an atom bomb.[5] The volume of chemical explosives onboard the *Mont-Blanc* far exceeded the anticipated volume of U235 or plutonium in an atom bomb. But U235 and plutonium would carry a far greater punch. The chemical explosive TNT yielded about 0.5 electron volts per atom. The split uranium nuclei in an atom bomb would yield two hundred million electron volts per atom.

CHAPTER TWENTY-TWO

CONSTRUCTION

With the nation ill-prepared for war, countless tons of steel, copper, concrete, and other supplies that would otherwise have been used to build ships, planes, and tanks was now being diverted to build the atom bomb. President Harry S. Truman would later call that decision "the greatest scientific gamble in history."[1]

Ever since Archimedes designed the first catapult for use against the Romans, warring nations had looked to scientists and engineers to build even more lethal weapons. The atom bomb would be in a class by itself. If it worked, if it could be assembled in time to use during the war, it would be the greatest leap in destructive power in the history of warfare. Armed with such a weapon, a single bomber could strike a blow that previously would have required 2,000 bombers dropping more than 150,000 500-pound bombs.[2]

The United States had gotten off to a late start, but it was now fully committed to beating the Germans. The nation that won the race could not only win the war; it could reverse a defeat. In the longer run, if Hitler obtained an atom bomb, he could extend the scourge of Nazism—genocide, forced sterilization, euthanasia—across the globe. Every Jewish man, woman, and child would be put to death.

In the spring of 1943, the hills and valleys of eastern Tennessee echoed with the sound of bulldozers and jackhammers clearing land where Tennessee hill folk once planted tobacco, raised hogs and chickens, and occasionally made a little moonshine on the side. Flatbed trucks carrying building material for houses clogged the country roads, and workers laid track for the railcars that would haul uranium ore and equipment to the giant U235 separation plants. The first equipment to arrive, some 2,157 carloads of electrical equipment, 1,219 of heavy equipment, 5,389 of lumber, 1407 of pipe and fittings, 1,188 of steel, and more would go to the site of Lawrence's future electromagnetic separation plant.[3]

As with everything General Groves oversaw, speed was essential. To work around chronic wartime shortages, construction companies used prefabricated panels made of "cemestos," a mix of cement and asbestos, for walls instead of wood, and cement blocks instead of poured concrete for foundations.

Over the next two years, eighty thousand acres of Tennessee farm country would give way to the towns of Oak Ridge and Happy Valley, home to some seventy-five thousand workers who would build and run the uranium separation plants.[4]

After months of practice, construction crews were finishing houses at a blistering pace of one every two hours, while work on the two U235 separation plants—each a unique design based on experimental technology—was constantly slowed by unforeseen glitches.[5] In October and November, trial runs of the electromagnetic separation plant repeatedly short-circuited, and equipment broke down and had to be sent back to the manufacturer for repairs.

Despite all the setbacks, the electromagnetic plant was still well ahead of the problem-plagued gaseous diffusion separation plant. After working six- and seven-day workweeks, engineers were still searching for ways to manufacture the first-of-a-kind airtight pipes and pumps and, most importantly, the porous metal tubes that would strain out the U238 atoms.

Betting that the problems would be solved, and unwilling to waste a single day, Groves had ordered Kellex, the company building the diffusion separation plant, to begin construction on the building's exterior even before the technology for the machinery inside was perfected. On October 20, 1943, crews began clearing and preparing the land for the enormous foundation. Meanwhile, a few thousand miles northwest of Oak Ridge, on a desolate tract of country half the size of Rhode Island, surveyors raced to map a site for the manufacture of plutonium.

The new town of Hanford, Washington, would support three nuclear reactors, chemical plants to separate the plutonium from uranium, and the forty thousand workers who would build and operate them. Managers at Hanford had one advantage over their colleagues at Oak Ridge: much of the technology for manufacturing plutonium and separating it from uranium was tested and proven. Still, Hanford had its own unique obstacles.

Many of the farmers refused to sell their land or insisted that they be allowed to stay through harvest time. The Corps of Engineers couldn't tell the farmers why they needed the acreage, and their secrecy gave rise to rumors that the whole business was a government land grab.

The farmers' protests were heard in Washington, D.C. where the Senate Committee on Military Spending threatened more delays. Groves and the committee chairman, Senator Harry S. Truman, had tangled frequently before the war when Groves was supervising military construction. Truman was preparing to investigate the farmers' complaints when he got a phone call from the secretary of war, Henry Stimson. Stimson warned him that an investigation would endanger national security, and the Senator agreed to postpone any inquiry until after the war.

The Corps of Engineers ended up condemning the land of the farmers who refused to sell, only to have to fight them in court. More time was lost. In all, some 1,200 properties were involved. Local courts, which could manage only about seven condemnation cases a month, were overwhelmed. Groves did everything he could to speed up the process, arranging for the Justice Department to bring in additional judges and empowering the department's attorneys to increase appraised values, thus encouraging out-of-court settlements. Unfortunately, the suits attracted the attention of local reporters who wrote stories that threatened to compromise security. At war's end, two hundred properties remained in dispute.[6]

There was also the issue of labor shortages. While Oak Ridge could recruit from the nearby city of Knoxville, the War Manpower Commission barred Du Pont and the other Hanford contractors from competing with the Seattle-based airplane manufacturer Boeing. Consequently, Du Pont's agents had to travel to the Great Plains to find workers.

Recruitment ads for Oak Ridge read, "When you're a grandmother, you'll brag about having worked at Tennessee Eastman."[7] Hanford's advertising cautioned that the area was "a little on the rugged side" and suggested recruits bring a padlock. Many of those who answered the Hanford call expected evergreen forests, snow-capped mountain peaks, and excellent fishing and hunting. What they got was desert, sagebrush, and jack rabbits. "It was so darned bleak," wrote one worker, "if I'd had the price of a ticket, I wouldn't have stayed."[8]

To make matters worse, construction had torn up much of the grass that anchored the sandy soil. Dust storms were so severe that buses had to turn on their lights during the day. In a big blow, the fine dust darkened faces and found its way into clothing and food. The stinging winds came to be known as "termination winds" because the number of workers booking tickets home always increased afterwards.

Groves pushed management to come up with perks that would encourage workers to stay. The food was good and plentiful despite wartime rationing. There were saloons, slot machines, a bowling alley, and movie theaters. The entertainment hall offered boxing on Monday nights and a name band Thursday, Friday, and Saturday evenings.[9] But despite Du Pont's best efforts, the men often found themselves with nothing to do after work. Boredom, loneliness, and drunkenness proved a combustible mix. There were frequent fights and occasional homicides, prompting Hanford managers to build their own jail and Du Pont to build saloons with windows that opened inward, making it easier to lob in tear gas.[10]

General Groves took Oppenheimer's suggestion. He purchased a boy's boarding school, the Los Alamos Ranch School, in the New Mexico desert about thirty miles from Santa Fe to serve as the principal laboratory for Oppenheimer's scientists. The school's location, isolated atop a mesa and accessible by a single winding dirt road, afforded excellent security for the top-secret operation. But in the spring of 1943, security was the least of Los Alamos's problems. The melting snow had left the roads ankle-deep in mud. Army Corps engineers were building tin-roofed plywood barracks for the scientists, but they weren't ready yet.[11] Most of them had to stay in dude ranches forty miles away. Laboratory equipment was sitting in crates waiting to be assembled. There was little food and only a single poorly functioning telephone line to the outside. But the most serious problem was not facilities; it was a lack of leadership.

When John Manley and Robert Wilson, two of Oppenheimer's early hires, arrived at Los Alamos, they were appalled at how disorganized everything seemed. Oppenheimer was still living in California at that time in his opulent home overlooking San Francisco Bay and was hosting a lavish cocktail party when the two men came to report what they had found. The Los Alamos scientific director was in no mood to hear their complaints. "Oppenheimer became extremely angry," Wilson later recalled. "He began to use vile language . . . [and said] it was none of our business."[12] Frightened by Oppenheimer's tantrum, Wilson and Manley decided they had to take more initiative and not rely so much on Oppenheimer.

Manley and Wilson weren't the only ones concerned about Oppenheimer's management skills. In the spring of 1943, General Groves assigned experienced Corps of Engineers officers to take charge of the planning and actual operation of Los Alamos.[13]

While Oppenheimer was adjusting to his new wartime responsibilities, Sam Goudsmit was learning a new job editing a handbook that explained his MIT radar project to generals. "I had to work with lawyers, engineers, administrators, security executives . . . and a charming husband-and-wife team of artists who were on leave from the Walt Disney studios," he told his wife Jean. But then, in the summer of 1943 after two years in Cambridge, Goudsmit was assigned a job that took him closer to the action. He was sent to England to help sort out a curious radar-related problem. RAF crews were pleased with their radar. American crews who used basically the same equipment were not. To find out about the discrepancy, Goudsmit interviewed dozens of flight crews. He lived with them, went on leave with them, and drank with them after hours, and eventually the answer emerged. The radar was best suited for low-flying tactical aircraft like the ones the RAF flew. The American Eighth Air Force flew strategic bombers at high altitudes.[14] Armed with that information, U.S. Army Air Forces shifted the radar to its low-flying tactical aircraft.

When Goudsmit returned to the United States, he came to the attention of General Groves as a possible candidate to head a new scientific intelligence unit he was forming. Goudsmit had a top security clearance, an expertise in nuclear physics, and personal relationships with Werner Heisenberg and several top German physicists. But it was his ignorance concerning the Manhattan Project that sealed the deal. If captured, Goudsmit could not yield any valuable intelligence.

CHAPTER TWENTY-THREE

HEAVY WATER II

While scientists and engineers in the United States raced to build the Allied bomb, a team of British and Norwegian commandos trained for one more secret mission to scuttle the Axis bomb program. Four months had passed since the failed mission to sabotage the Norwegian heavy water plant. In that time the Germans had beefed up security, adding more troops, machine guns, searchlights, and even a perimeter minefield. The British had upped their efforts as well.

Two Norwegian engineers, Jomar Brun and Einar Skinnerland, who had designed the Norsk Hydro plant, were now working with the Allies in England. Armed with the plant blueprints, they constructed a mockup of the electrolysis area where the heavy water was distilled so that six Norwegian saboteurs could practice laying dummy explosive charges in the dark. This time there would be no gliders. The advance team that had been in place, living off the land since the first ill-fated attempt, would mark the drop zone, and the sappers would parachute in.

On the first attempt, the plane carrying the sappers could not find the advance team's beacon and had to return to Scotland. On a moonless night weeks later, their luck changed. They parachuted onto a snow-packed mountain plateau, a barren landscape inhabited only by reindeer. Fighting gale-force winds, they managed to find the supplies they had also parachuted in and meet up with the advance team.[1] After days of skiing, the whole group finally sighted their target.

The imposing seven-story Norsk Hydro plant sat on a ledge at the bottom of a steep bluff. Alongside, several large pipes carried water from the lake above to the gorge below. The saboteurs' intelligence briefs had been spot-on. The guards, the machine guns, and the rest of their security measures were exactly where they expected.

Just after the 10:00 p.m. shift change, the eight men began their assault. In a heroic feat of mountaineering, they eluded the German guards by descending into the gorge, crossing a frozen river, and climbing the five-hundred-foot rock face on the other side. From there, they cut through a wire fence and crawled through a cable-intake opening into the plant. There the demolition team went to work, laying explosive charges, setting the fuses, and retracing their steps. They had less than a minute to get away before the explosives detonated.[2] Miraculously, all the men escaped uninjured. British intelligence estimated it would take twelve months for the Germans to get the plant up and running again.

General Groves didn't learn about the Norsk Hydro raid until two weeks later when a Swedish newspaper reported it. The British had informed him months earlier that they were planning a commando operation but never updated him. Groves was predictably furious.

Channels of communication between the U.S. and British intelligence would remain strained during much of the war. The problems dated back to 1941 when British scientists had the technological edge over American scientists and rejected U.S. efforts at close cooperation. In 1943 when American scientists took the lead, Groves turned the tables and locked the British out.[3] "It became quite evident," he would later write, "that their primary purpose [in cooperating] was to follow our work and to obtain information from us."[4] But Groves's exclusion of British scientists came at a price. In response, the Brits withheld intelligence and kept secret the existence of an important source: a British spy in the upper echelons of Germany's scientific and academic community.[5]

CHAPTER TWENTY-FOUR

LOS ALAMOS

The new recruits began arriving at the small train platform in Lamy, New Mexico, in April 1943, four years after the discovery of atomic energy. There were a few families, but single men, generally postgraduate students, predominated. Many came from MIT where they were working on the top-secret program to develop airborne radar. Oppenheimer knew he would have a hard time recruiting them. They were already doing important work, and he couldn't reveal what he wanted them for. He could only offer his standard pitch: if successful, he told them, they could save American lives. Instead, he drew on his personal relationships with the radar program's senior scientists, many of whom he had known for years. He convinced them to come on board, and their postgrads soon followed.[1]

From Lamy, the recruits made their way to 109 East Palace Avenue in Santa Fe, where Dorothy McKibben, a forty-five-year-old widow hired by Oppenheimer, provided the newcomers with passes and arranged transportation to the Hill, a secret site on a mesa high above the desert floor. Los Alamos, the hub for America's atom bomb program, would eventually grow to several thousand scientists and support personnel.

With Oppenheimer looking on, each new group of scientists filed into a large wood-paneled room that had served as the reading room for the Los Alamos Ranch School. The school had issued its last diplomas at the end of a shortened school year just four months earlier. Now called Project Y, the top-secret Los Alamos laboratory did not appear on any maps. To the U.S. Postal Service, it was simply P.O. Box 1663, Santa Fe, New Mexico.

Inside, Groves greeted the new arrivals in a manner that could charitably be described as clueless. Robert Wilson, a member of the audience, remembers, "He appeared not to believe in the eventual success of the project. He emphasized that if—or when—we failed, it could be he who would have to stand before a congressional committee to explain how money had been squandered."[2] The young scientists who had put their lives on hold and traveled across the country to the middle of nowhere to answer their country's call had to wonder whether they had made a terrible mistake.

But then, Robert Serber, Oppenheimer's associate from U.C. Berkeley, stood up to speak. Shouting to make himself heard over the noise of hammering and sawing, he began, "The object of the project is to produce a practical military weapon in the form of a bomb in which the energy is released by a fast neutron chain reaction."[3] The young recruits looked at each other. They'd been told that their work might shorten the war. Few, if any, had imagined they would be building an atom bomb.

They would design two types of bombs, Serber explained: one fueled by U235, the other by plutonium. How much U235 or plutonium was needed was yet to be determined. They were still working to figure out the critical mass—the minimum amount of fuel required to trigger a sustained nuclear chain reaction—for each bomb. Preliminary estimates put the amount of U235 at thirty-three pounds (that number would later quadruple), and the plutonium critical mass at eleven pounds.

Serber then moved on to the issue of how to detonate the bomb—a question they were still working on that required additional research. Oppenheimer's team planned to use the same gun-type detonator for both bombs. The detonator would fire a projectile of plutonium or U235 down a cannon barrel at a target made of that same plutonium or U235. The two would merge, forming a critical mass which would trigger a nuclear chain reaction and an explosion.[4]

In his Chicago nuclear reactor, Enrico Fermi had slowed neutrons down to generate a controlled chain reaction, but slow neutrons would not work in a bomb. For a bomb, Oppenheimer needed super-fast, turbo-charged neutrons, and plenty of them. Oppenheimer, Fermi, and colleagues had done extensive research on slow neutrons, but they knew relatively little about fast neutrons. They still needed answers to basic questions like how many neutrons a U235 atom gave off when its nucleus was split by a fast-moving neutron.[5]

As he raced to catch up to Heisenberg, Los Alamos's scientific director knew there was no margin for error. The difference between success and failure—and perhaps the outcome of the war—would be determined in roughly a millionth of a second, the time it would take to split eighty generations of uranium nuclei.

The formation of the critical mass would trigger a frenzy of careening neutrons. Twenty, then thirty and forty generations of nuclei would split, producing energy and generating heat of a million, then two million degrees Fahrenheit. Pressure inside the bomb would build. The chain reaction would continue, splitting fifty, then sixty and seventy generations. The heat would intensify: ten million, then twenty million degrees Fahrenheit.[6] And then, if everything went as designed, the bomb would explode.

But if things did not go as planned, if the swelling explosion drove the atoms in the critical mass apart before an appreciable fraction of the atoms had fissioned, neutrons would escape through the gaps, the chain reaction would peter out, and the searing heat, now equivalent to that in the interior of the sun, would incinerate the bomb.[7] There would be an explosion, more like a fizzle, equivalent to several tons of TNT, but nothing comparable to an atomic explosion.

The challenges were staggering, and Oppenheimer got off to a shaky start. Directing the bomb project required far more than a brilliant mind. It took extraordinary stamina, patience, coolness under fire, and an ability to manage and motivate others. Oppenheimer had already proven himself a poor manager. In his first weeks at Los Alamos, he did little to inspire confidence. "He was such a smart aleck," commented Robert Wilson, a former student, "and didn't suffer fools gladly."[8] But it was his politics that raised the most serious concern.

In June 1943 while in Berkeley on business, Oppenheimer got in touch with an old flame, Dr. Jean Tatlock, a pediatric psychiatrist and a former member of the Communist Party. She was unhappy and had written several letters begging to see him. Oppenheimer must have known that Colonel Pash's men would be watching, but he arranged a rendezvous, nonetheless. Every move was observed and duly reported: "The two had drinks. 10:50 p.m., Oppenheimer and Tatlock drove back to her apartment. 11:30 p.m., the lights went out. 8:30 a.m., Oppenheimer exited the building.[9]

Two weeks after that meeting, Colonel Boris Pash wrote his superiors at the Pentagon recommending that Oppenheimer be denied his security clearance and fired.

By the summer of '43, the stress of the Tatlock affair, the constant surveillance, and the six-day workweeks had taken a toll at work and at home. Oppenheimer began pinning top-secret documents in his pockets to ensure that he didn't misplace or lose them. His wife Kitty had not adapted well to the fenced-in life at Los Alamos and was drinking heavily. Depressed and exhausted, Oppenheimer confided to a friend that he was at the end of his rope and close to quitting. The friend's response: "There isn't anybody else who can do it."[10]

Depression was a serious problem for Oppenheimer. At Cambridge, where he studied before coming to Göttingen, he had done poorly in experimental physics. Conducting laboratory experiments and meticulously recording the results was a "terrible bore," he wrote to a friend, "and I am so bad at it that it is impossible to feel that I am learning anything." Oppenheimer was so depressed he reportedly plotted to punish his experimental physics tutor by leaving a poisoned apple on his desk, hoping he would take a bite. It's not clear whether he actually carried out the plot. In any case, the tutor was not harmed. When university administrators learned about what Oppenheimer had planned, they voted to expel him. But his parents, who were visiting Cambridge at the time, lobbied the university to put their son on probation instead of expulsion. The university agreed on condition that he undergo regular sessions with a prominent London-based psychiatrist. The incident occurred in the autumn of 1925, and over the next several months, Oppenheimer saw at least three psychoanalysts.[11]

Ironically, it was theoretical physics that ultimately helped him heal. Oppenheimer read several papers in the emerging field of quantum mechanics—including one by a young German, Werner Heisenberg—found new direction, and enrolled at Göttingen. There he regained his intellectual footing. Too much so at times. Eager to demonstrate how smart he was, he often interrupted his fellow students, prompting them to sign a petition threatening to boycott the class if the professor didn't rein in the child prodigy. Unwilling to confront him directly, the professor called Oppenheimer to his office on another matter and arranged to leave the petition on his desk where Oppenheimer was sure to see it while he stepped out for a few minutes. The ruse worked. Oppenheimer quieted down after that.[12]

Years later at Los Alamos, it was General Groves's vote of confidence that helped restore Oppenheimer's faith in himself. Groves decided Oppenheimer was essential to the project, and they couldn't afford to lose him. His mind made up, Groves brushed aside Pash's recommendation and granted Oppenheimer his security clearance.

With the security issue settled, Oppenheimer's spirits improved, and he soon made clear he was the right man for the job. "He knew and understood everything that went on in the laboratory whether it was chemistry or theoretical physics or machine shop," wrote Manhattan Project scientist Hans Bethe.[13] And Oppenheimer wanted others to be similarly informed.

Early on, General Groves had imposed strict security measures, insisting that information be compartmentalized, that the chemists, metallurgists, theoretical physicists, and others be told only what they needed to know to do their jobs. Oppenheimer objected, arguing that he needed everyone contributing and sharing their ideas. In a rare retreat, General Groves acquiesced.

The Manhattan Project changed Oppenheimer, making him more empathetic. It also taught him how to lead and how to manage the different, often difficult personalities on his team—among them, Richard Feynman, a brilliant, eccentric, and stubbornly independent theoretical physicist. Oppenheimer knew Feynman would bridle at working on a highly regimented, military-run project. He also knew that Feynman's wife was seriously ill with tuberculosis, and that Feynman would never leave her. The canny Oppenheimer pulled some strings and managed to secure a wartime train pass for her to come to New Mexico. He also found her a bed in a hospital close to Los Alamos where Feynman could visit regularly, and Feynman agreed to sign on.[14]

In time, Oppenheimer grew to be a decisive, even inspiring leader known simply as "Oppie" to everyone on the mesa. Doubters became ardent supporters. "When I was with him, I was a larger person," said Robert Wilson, who had previously called him a smart aleck. "I became very much of an Oppenheimer person and just idolized him . . . I changed around completely."[15]

"He never needed to show off or shout," commented a young telephone operator. "He could have demanded priority one with his telephone calls but never did. He never really needed to be as kind as he was."[16]

The mantle of leadership had improved Oppenheimer's social skills, but he could still lapse into his former self—the absent-minded professor. One day, lost in thought, he drove up to the main Los Alamos security gate and failed to stop. The startled MP shouted a warning and then fired at the car's tires. After that, Oppie was not allowed to drive himself. From that day forward, General Groves insisted he have a military driver.[17]

CHAPTER TWENTY-FIVE

NOTHING BUT THEORY

Oppenheimer's Manhattan Project included eight Nobel laureates and more than a dozen future laureates.[1] Their mission would require every bit of that intellectual horsepower. During its first year of operation, Los Alamos had only a single gram of purified U235, and even less plutonium. The samples were too small and too valuable to risk losing during an experiment, so all the initial research was strictly theoretical.[2]

The explosion would progress in stages. The U235 core and the tamper, a neutron reflector surrounding the core, would start out as metal and change instantly as the metal heated from solid to liquid to gas. Project scientists had to calculate how the explosion would develop at each of those stages and how long it would take to split eighty generations of atoms.[3] The answer to that question begged more questions. An explosion required a critical mass of either U235 or plutonium. How much was a critical mass? And how pure did the plutonium and U235 have to be?

In peace time, under normal conditions, engineers would have waited until they knew the answers before deciding how many nuclear reactors to build at Hanford and how many uranium separation plants to build at Oak Ridge. A five-hundred-pound critical mass would require a much larger bomb casing and far more processing capacity than a fifty-pound critical mass, a lot more electricity, and a larger power plant. The size of the critical mass would factor into manpower decisions: how many engineers, carpenters, pipefitters, and electricians to hire; the number of worker houses, barracks, hospitals, and grocery stores to build. But these were not normal times. The United States was fighting a war. The

Germans were also building an atom bomb, and they had a two-year head start on the United States.

In his autobiography Groves explained, "With time as the controlling factor we could not afford to wait to be sure of anything."[4] Instead, they rushed ahead. Based on that one gram of U235, a microscopic speck of plutonium, and some educated guesses, they set about building two massive industrial complexes. Redundancy was their best hedge against failure. If the U235 bomb proved impossible to build, they would have the plutonium as back up. If the electromagnetic separation plant didn't work, they would have the gaseous diffusion plant. The industrial complexes built to manufacture the U235 and plutonium were enormous—so large, they rivaled the size of all the country's automobile manufacturing plants put together.[5]

Speed was not just a Groves obsession; it was a necessity. Everything associated with the Manhattan Project had to move at an accelerated pace. The scientists worked day and night to shave fractions of a millionth of a second off the estimated chain reaction time. They surrounded the critical mass with a reflector to bounce escaped neutrons back into the mass. They added a supplemental source of neutrons, an initiator, that sprayed additional neutrons into the critical mass.

But no matter how much progress the Los Alamos scientists made, no matter how many research hours they put in, they couldn't shake the fear that they were losing ground to their nemesis Heisenberg.[6] Werner Heisenberg never set foot on the Los Alamos mesa, yet Hitler's chief atomic scientist was a constant presence.

One afternoon, after workers installed a public address system in the top-secret technical area, an anonymous caller asked the Los Alamos operator to page one of the scientists. The operator tried two or three times and was puzzled when each announcement drew bursts of laughter. "Dr. Werner Heisenberg, please report to the office. Dr. Heisenberg please report to the office."[7]

CHAPTER TWENTY-SIX

DUMB LUCK

General Groves and Oppenheimer had attracted some of the world's most brilliant scientists to work on the Manhattan Project. But to succeed, they needed more than just sharp minds. They needed to be lucky. They needed the kind of luck that brought Edward Norris, an interior decorator in Greenwich, Connecticut, to Columbia University's physics department.

Norris's work occasionally required him to do some painting. He preferred using a spray gun, but the gun didn't apply the paint evenly. It was always thicker in some spots than others. So Norris invented a device that solved the problem by spraying the paint through a fine metal mesh, which he invented himself using a process called electrodeposition. The device worked so well, he was able to sell it to a manufacturer of woven wire, which hired him to make the "Lectromesh."

Norris's invention became known in Britain as well, and in the summer of 1941, a British scientist approached him with an unusual request for "a half-inch square screen with a half-million holes in it."[1] When the job was completed, that scientist introduced Norris to Dr. Dunning at Columbia University, who was working on some sort of secret government project. They wouldn't tell Norris what it was, but he remembered reading an article in the *Saturday Evening Post* that mentioned something called gaseous diffusion. Norris found the article in his local library, figured out the secret, and set about inventing a more suitable screen, which he brought to Dunning at Columbia, and the former interior decorator was eventually hired.

On a Saturday morning in March 1943, Elky Shazkin, a young physicist, came to Columbia University's physics building accompanied by her husband Leonard Shazkin to turn down a job offer on a top-secret government project. She had been told it would involve long hours. Her husband had just been drafted into the army, and she wanted to spend time with him before he left.

Her prospective boss, professor Dr. Francis Slack, expressed disappointment, and as Elky was leaving, asked what her husband did for a living. She told him he was a production manager and printing expert at *House Beautiful* magazine and had a BS from Carnegie Tech in printing.

Slack suddenly got very excited. When Elky mentioned her husband was in the hall waiting, Slack asked him to come in and called to an associate, Edward Adler, to join them. The two then spent the next half hour bombarding Leonard Shazkin with technical questions about printing and the type of printer he used. Bewildered, Shazkin answered their questions and was shocked when they offered him a job and asked him to start the next day. Shazkin explained that it was impossible because he had to report for basic training, and they told him not to worry, they would take care of that.[2]

A short while later, after completing a survey of industrial printing machines in the New York City area, Adler and an army captain showed up at the American Chicle Company (makers of Chiclets chewing gum) and informed the company's president that they needed to commandeer the company's five-color Chambord press, which Chicle used to make display packages. When the surprised president asked who they were and what they needed the press for, the army captain told him it was for a top-priority war project and apologized because he couldn't say any more than that.

For several months thereafter, Adler and Shazkin became frequent visitors to Chicle's headquarters where they spent hours printing dots of a special shape, viewing them under a microscope, making adjustments, and printing some more. Each day, when the two left, a detachment of MPs swooped in to dust off the press and remove all evidence of their work.[3]

After many hours of trial and error, Adler and Norris invented a prototype for a thin metal filter with millions of holes per square inch that was flexible enough to shape into a tube and could resist highly corrosive uranium hexafluoride gas. On April 1, 1943, General Groves gave the Houdaille-Hershey Corporation the go-ahead to make plans for a new factory to manufacture the Adler–Norris filter in Decatur, Illinois. And, on May 30, 1943, construction began on the gaseous diffusion plant at Oak Ridge.

CHAPTER TWENTY-SEVEN

HUNTING HEISENBERG

BBC RADIO, FEBRUARY 9, 1943, ROBERT ROBINSON REPORTS

The triumphant conclusion of the battle of Stalingrad and the capture of eight more German generals and forty-five thousand other prisoners in the past two days has overshadowed the rest of the news from Russia. Our allies have kept up their great advance on the southern front.

By the spring of 1943, the momentum of the war in Europe had shifted in the Allies' favor. The German Sixth Army in Stalingrad and the vaunted Afrika Korps commanded by General Erwin Rommel in North Africa had both been defeated. The depressing war news only added to the gloom many were feeling on the German home front where shortages of food, gasoline, heating oil, and other essentials were widespread, and Allied bombing raids on Berlin and Leipzig were becoming more frequent.

On March 1, 1943, Heisenberg had just attended a lecture at the Air Ministry in Berlin when the air raid siren went off. He and the other attendees rushed to the building's bomb shelter as hundreds of Allied bombs pounded the city, collapsing parts of buildings and setting houses on fire. He survived but was terribly worried about his in-laws who lived outside the city and two of his children who had come to celebrate their grandfather's seventy-fifth birthday. Heisenberg hiked for an hour and a half through rows of burning houses and smoldering piles of rubble to get to their house. Upon arriving, he discovered that the doors and shutters had been blown out and his in-laws and children were nowhere in sight. Racing through what was left of the front door and up to the attic, he found his mother-in-law desperately trying to beat out the flames of burning timbers. She told him everyone was safe. The children had been taken in by a neighbor.[1]

A few weeks later, Heisenberg moved his family to their mountain cottage in the town of Urfeld in southern Germany while he stayed behind to supervise construction of his largest nuclear reactor in the newly constructed underground bunker in Berlin.[2]

At the same time, in Chicago and Los Alamos, Oppenheimer's European émigré scientists were growing more and more anxious, fearing that the bombing raids on German cities and the reversals suffered by the German armies in North Africa and Stalingrad would make Hitler desperate.[3] Newspaper articles quoting a rousing speech by Hitler's propaganda minister Joseph Goebbels did nothing to calm their rattled nerves: "I ask you: Do you want total war? Do you want it, if need be, even more total and radical than we are capable of imagining it today?"[4]

Rumors spread that the Germans would have radiological weapons ready to go by the end of 1942 and that the army was setting up radiation detectors all around Chicago, and some scientists were moving their families out of the city.[5] One scientist urged General Groves to "warn the American people in an official broadcast that the United States might be hit by an atomic bomb."[6] Groves declined, but months later elected to send a different message—an air mail special delivery from the combined RAF and USAAF Eighth Air Force.

On December 3, 1943, a wave of British long-range Lancaster bombers swept across the skies above Leipzig, Germany. Their target: the Institute of Theoretical Physics where Werner Heisenberg was reportedly working. RAF bombs destroyed the institute's top floors and Heisenberg's scientific papers along with them. They also damaged Heisenberg's home and the garden that Elisabeth had spent years cultivating. But they failed in their primary mission: "the killing of scientific personnel employed therein."[7]

Allied bombers followed the Leipzig attack with a raid on Otto Hahn's chemistry lab outside Berlin, but Hahn also escaped injury. By late 1943, it had become obvious that the bombing raids, the power outages, and other disruptions made it impossible to build the next generation *B-VI* reactor in the Berlin bunker, so officials began relocating personnel to new facilities in southern Germany. Up to that point, General Groves had been receiving regular intelligence reports on Germany's uranium research program. The relocation meant that the intelligence virtually ceased.[8]

CHAPTER TWENTY-EIGHT

PEENEMÜNDE

Following the Leipzig and Berlin bombing raids, a rash of rumors about Hitler's long-promised "wonder weapons" spread through the cafés and taverns where Germans gathered after work. There was talk of "giant guns" and "stratosphere guns and rockets" that would soon enter the battle.[1] After a year of defeats in North Africa and Stalingrad and the bombing of German cities, the demoralized German public at last had some good news. To boost public morale, the security police leaked a list of super weapons that they said were being developed.

Word of the "wonder weapons" eventually made its way to London and the scientific section of MI6. With Hitler's armies on the defensive and Heisenberg's disappearance stoking fears of a German atom bomb, one rumor drew special scrutiny. It concerned a rocket armed with "explosives of atom-splitting type."[2]

Britain's military intelligence had been uncovering clues of a secret rocket program ever since the launch of the war. In November 1939, an anonymous "well-wishing German scientist" sent an unsolicited package to the British Embassy in Oslo, Norway. The so-called Oslo Report contained descriptions of several technologically sophisticated weapons that the writer said were under development; among them, a rocket-powered flying bomb. British intelligence believed the report was German propaganda, and it was mostly ignored.

Four years later, hidden microphones at a POW camp for high-ranking German prisoners north of London picked up a conversation between General Ludwig Cruewell, deputy commander of the German Afrika Korps, and Ritter von Thoma, another Afrika Korps general.[3] The two men knew each other, so the British housed them in the same cell, hoping they might talk freely. And talk they did. Von Thoma expressed surprise that they had not heard explosions coming from the direction of London. He'd been told the city would be targeted by a barrage of newly developed pilotless flying bombs.[4]

The mention of London and flying bombs suggested a possible explanation for several ski jump–like structures the Germans had built along the French coast. Forty feet long with an upward curve at the end, they were pointed in the direction of London.[5] Were they launch pads for pilotless flying bombs? The Treaty of Versailles which ended World War I had placed strict limits on the number of conventional weapons the Germans were allowed. Had they invented a new unimagined weapon to circumvent those limits?[6]

Evidence of a program to build pilotless flying bombs mounted in June 1943 when two Luxembourgers conscripted by the Germans to work as forced labor in Pennemünde, a German municipality on the Baltic Sea, managed to get word out that they were building large rockets with powerful engines that made a deafening roar.[7]

None of the intelligence mentioned rockets with "explosives of atom-splitting type." But then a British analyst discovered that Pascual Jordan, a theoretical physicist, worked at Peenemünde. Jordan had studied in Göttingen and co-authored several scientific papers with Heisenberg. The most alarming piece of intelligence came from the American Office of Strategic Services (OSS), predecessor to the CIA. A Frenchman who had escaped from a German work detail told the OSS that his job was guarding barrels of water being shipped from Rjukan in Norway to Pennemünde. The OSS agent who received the report had attended a seminar on nuclear fission the week before. He suspected the barrels contained heavy water, so he pulled out some maps and an encyclopedia and discovered that Rjukan had a large hydroelectric plant manufacturing heavy water.[8]

Spurred by the mounting intelligence, the RAF began regular reconnaissance flights over the area. Their photographs revealed a large industrial complex, but no signs of a rocket. Then, on June 28, one of the reconnaissance photos captured a blurry image of a "whiteish cylinder about 35 feet long and 5 or so feet in diameter with a bluntish nose and fins at the other end."[9] It was the first Allied sighting of Hitler's secret vengeance weapon, the V-1 flying bomb. Determined to strike before the bombs could be fired, British Bomber Command recommended a nighttime raid by several hundred bombers. But because it was summer, the long days meant there were not enough hours of darkness for such a massive raid, and they had to wait.

Finally, on the night of August 17 and into the morning of August 18, six hundred RAF aircrafts released their bomb loads over the secret site. Bomber Command had opted to target the homes of the scientists and engineers who built the rockets rather than the catapults and rockets themselves. Unfortunately, a substantial portion of the bombs missed their mark and struck the camp of the foreign workers. Bomber Command estimated that the raid delayed the V-1 program from four weeks to six months.[10]

CHAPTER TWENTY-NINE

HAMBURG

> **CBS NEWS, JULY 25, 1943, EDWARD R. MURROW REPORTING**
>
> Last night the Royal Air Force went to Hamburg, Germany's largest port and main center of their submarine building industry. The British bombers dropped more than 2,300 tons of bombs. It was the greatest weight of bombs ever dropped in any raid.

The decision to target the scientists and engineers at Pennemünde came on the heels of a combined attack by the RAF and the Eighth USAAF and a change in Allied bombing tactics. After two years of war, during which thousands of British civilians had been killed and large swaths of British real estate burned to the ground, the Germans had taught the RAF a valuable lesson which they put to use in Hamburg.

The RAF flew nighttime raids during which they firebombed high-density commercial and residential areas of the city, killing workers and disrupting civilian life. The United States, however, pursued a strategy designed to limit civilian casualties. Called the "precision bombing doctrine," it required U.S. Army Air Force to use high-explosive bombs instead of incendiaries and to target strategic objectives like weapons factories and airfields. Targeting civilians was strictly prohibited. The raids were to be conducted in daylight from a high altitude out of range of enemy anti-aircraft fire. The bombardiers were required to have eyes on their targets before releasing the bombs. If the targets were obscured by clouds or smoke, the crews were not permitted to release their bombs. With the bombs onboard, they had to fight their way through flak and enemy fighters and make their way back to base.[1] The USAAF bent those rules slightly in Hamburg by dropping incendiaries during the day to mark targets for the RAF's nighttime raids.

The precision bombing doctrine had been adopted in response to the carnage of World War I, during which some thirty-seven million people were killed or injured. It was rooted in the belief that the USAAF's newly invented, highly touted Norden bombsight would make U.S. bombing more accurate. It could therefore win wars and save lives by destroying the aircraft and tank factories, railroad yards, airfields, and oil refineries needed to fight the war rather than killing millions of soldiers and civilians.

The attack on Hamburg—eight days of intensive round-the-clock bombing, code-named Operation Gomorrah—began on a hot summer's night with low humidity. The night and day bombing exhausted the city's firefighting resources. Small fires merged to become large fires. Heat from the fires created a chimney effect. The rising hot air from the fire created an updraft that suctioned in the cooler surrounding air, generating a firestorm of hurricane-force winds that swirled around the fire-toppling buildings and trees and sucking human beings into the inferno.[2]

Neither the British nor the Americans admitted to targeting civilians.[3] Nevertheless, the firebombing of Hamburg resulted in a death toll of roughly thirty-seven thousand, the highest loss of civilian lives in one city during the whole European war.[4] "No city on earth ever felt such a shock of bombs as Hamburg experienced Saturday night," the *New York Times* reported.[5] The United States had maintained its official policy of only attacking strategic targets, but by assisting the RAF in bombing civilians, it had taken a step in eroding that policy. Bombing German non-combatants would ultimately pave the way for targeting Japanese non-combatants.

Two days after the bombing of Pearl Harbor President Roosevelt in one of his fireside chats warned the American people: "We must face the fact that modern warfare as conducted in the Nazi manner is a dirty business. We don't like it—we didn't want to get in it—but we are in it and we're going to fight it with everything we've got."[6]

In Chicago, while the rest of the country cheered the joint British and American raids on Germany's largest seaport, the Met Lab's émigré scientists once again held their breath, fearing that Hitler would not let such a deadly attack go unanswered. Rumors spread that he would counter with a grandiose and unexpected attack on the United States, perhaps on Chicago.[7] A Berlin radio broadcast on December 3, 1943, only intensified those fears when the announcer proclaimed that Germany would win the war in "one fell, drastic stroke."[8]

Scientists at the Met Lab in Chicago felt especially vulnerable. They were in a crowded city. Their Los Alamos colleagues were far more safely situated in the middle of the desert protected by barbed wire fencing, military checkpoints, and patrols. Rumors of an impending attack using radioactive poisons ran rampant. Coincidentally, Groves had commissioned a study of the German radiological threat and received its findings earlier that month. The author warned that it was extremely probable that the Germans had the means to conduct such an attack. He estimated that one week's output of fission products from a nuclear reactor spread over two square miles of a city like London would require that the entire area be evacuated, and a considerable proportion of the people would be incapacitated. The report concluded "if any attack is made," however, "it seems extremely likely that it will occur not in the United States, but in Great Britain."[9]

Not taking any chances, Groves ordered that Manhattan Project district offices in Boston, Chicago, New York, and San Francisco all be equipped with Geiger counters.[10] He worried about radioactive contaminants on the battlefield as well. The Germans could mix them with standard explosives and deliver them from the air in a bomb or a rocket, spray them from low-flying planes as a liquid or a dust, or bury them in land mines that they detonated remotely. Groves seriously considered that the Wehrmacht might lay down radioactive barriers to block the advance of Allied troops during the planned invasion of Europe.[11]

He was concerned enough to send one of his staff to London to brief the Allied Supreme Commander General Dwight Eisenhower. Groves's aide alerted the Army Medical Corps to look for unusual occurrences of fatigue or nausea and fogging of photographic or x-ray film, indicators of radioactivity. Portable Geiger counters were ordered, personnel trained in how to use them, and special equipment stockpiled in England and marked for immediate shipment should the need arise.[12]

CHAPTER THIRTY

THE KELLEX COMPANY IS HIRING

Clarence Johnson, the son of Swedish immigrants raised on a farm in western Canada, drove to school with his brothers and sister in a horse and buggy in warm weather and a sleigh in the cold of winter. While in high school, he trapped wild animals and sold their skins to raise money for college. A gifted and determined student, he eventually won a scholarship to study for a PhD in chemical engineering at MIT. And, while there, he learned that the Kellex Company in New York City needed engineers to work on some sort of war contract. He called and got an appointment for an interview.[1]

Johnson had a lame leg, and the day of his interview the elevator operator's union was on strike, so he had to climb twenty-three flights, arriving just in time for the interview. Dobie Keith, the man General Groves had hired to head the development of gaseous diffusion separation technology, had been impressed by Johnson's resume. Now, the determination he showed to make it all the way up to the office sealed the deal. Keith hired him on the spot. The twenty-nine-year-old chemist would soon rise to become head of the company's two-hundred-person laboratory.

On the evening of June 18, 1943, Johnson had just finished a frustrating series of tests, trying to combine the best features of the Adler-Norris filter developed at Columbia with a second promising design, but something always went wrong. The filter was too brittle, it wasn't sufficiently porous, or the holes were the wrong size. At 8:00 p.m., Johnson and his assistant decided to try one more experiment before heading home. This time when they bent the material, it didn't break. The two men could barely restrain themselves. But the material had to be not only flexible, it had to be extremely porous with billions of holes per square inch.

Two hours later, after several more tests, they had their proof. The material performed exactly as they had hoped. The next day, Johnson called Dobie Keith with the good news. "Do you think you can reproduce it?" Keith asked. "Come and see me immediately," Johnson answered.[2] After more than a year of experimenting, Kellex finally had an effective filter. But there remained an enormous challenge: manufacturing the filter on an industrial scale.

CHAPTER THIRTY-ONE

ALSOS

General Groves already had a full plate managing construction at Oak Ridge, Hanford, and Los Alamos. But in the fall of 1943, he was asked to take on another task: gathering intelligence on German atomic research. Accordingly, he set up a special unit called Alsos, comprised of military and scientific personnel.[1] Its mission: to track down evidence of Germany's atomic bomb program and seize control of their uranium supply. The unit would eventually grow to 201 members, including 71 scientists and the rest, military personnel.[2] He named the University of Michigan physicist Sam Goudsmit, Werner Heisenberg's old friend, to be its lead scientist.

Groves created the unit, in part, to calm the nerves of his émigré scientists.[3] Heisenberg had become such a distraction, explained one of the general's aides, that "Groves had difficulty keeping the scientists' minds on [their] work."[4]

Ironically, British scientists, those most likely to be targeted by Hitler's bombs, were less troubled. MI6's scientific intelligence unit had been sifting through mountains of data ever since the Battle of Britain, trying to assemble it into actionable intelligence. They knew that the Germans had accelerated heavy water production in Norway. But they had not discovered any nuclear reactors or giant uranium processing complexes like those at Hanford or Oak Ridge. Their source in Berlin, who had been monitoring Heisenberg's goings and comings, reported no mysterious travels or long absences from Berlin.[5] And a review of German scientific journals turned up several nuclear fission research papers that MI6 analysts believed would have been censored had the Germans been working on a bomb. With everything considered, British intelligence continued to maintain that Heisenberg was not a serious threat.

U.S. analysts disagreed. The British, they argued, had failed to account for fifty-one of the sixty German scientists known to be working in the field of nuclear fission. Allied intelligence had identified twenty-nine laboratories and thirty-one corporations working in support of Germany's atomic research program.[6] And the OSS had identified eleven papers on nuclear fission research, which they believed bore the fingerprints of the Nazi's Propaganda Ministry. The Germans were trying to "deceive us," Groves concluded. "We couldn't take the chance."[7]

Groves had another reason to be suspicious. Oppenheimer had proposed a similar ruse to confound the Nazis. He had even approached a colleague with "talents for physics and burlesque" and asked him to publish "innocent" articles under the names of Manhattan Project scientists.[8] That talented colleague preferred to spend his time doing legitimate research and respectfully declined.

Now his own intelligence chief, General Groves turned to Oppenheimer for advice. Oppenheimer did as British analysts had done two years earlier with their scientists. He gave General Groves a list of eminent scientists likely to be working on a German bomb and suggested where to look for the bomb-making and uranium-processing facilities. They would be located a long distance from the Soviet border and out of range of British bombers. Plutonium reactors would be "operating . . . where water [needed to cool the plants] is plentiful and where the flow from the plant passes either through open country or through country inhabited by an 'inferior race' whom they do not mind killing off."[9]

Armed with that information, U.S. and Royal Air Force reconnaissance flights prowled German skies looking for uranium and plutonium processing plants. Specially equipped A-26 Invader bombers sniffed the air for radioactive gases produced by nuclear reactors. And Allied agents sampled German rivers searching for radioactive water. These were dangerous missions. The A-26 crews often had to fly within range of German anti-aircraft fire. Weeks of reconnaissance turned up nothing, but Manhattan Project officials were still not convinced.

Maybe Heisenberg didn't need a large industrial plant like Oak Ridge to separate out U235. "[Heisenberg] might have come up with a way to do it in his kitchen," Oppenheimer warned.[10] Perhaps Heisenberg found a different way to produce plutonium and didn't need to build large nuclear reactors.

HEAVY WATER III

After the daring February 1943 commando raid, British intelligence estimated it would take the Germans a year to repair the Norsk Hydro plant. But it was up and running again after just six months—and producing more heavy water than before the raid: 6.6 kilograms per day instead of five.[1] Now the Allies planned a third attack.[2] This time it was an American operation.

On November 16, shortly before dawn, two hundred B-17s of the Eighth U.S. Army Air Force took off from a base in Great Britain. Almost two dozen were forced to turn back within the first hour due to mechanical problems.[3] The attack was planned to coincide with a lunch break when workers would be in a basement cafeteria and thus safe from the bombs. Unfortunately, the bombers arrived twenty-two minutes early. To kill time, the squadron commander headed back toward the North Sea, executing a wide 360-degree turn.

By the time they returned, the Germans were ready for them. One bomber was shot down by ground fire; another was hit, and the crew was forced to parachute into the sea.[4] One hundred seventy-six planes made it to the plant, however, and were followed by a squadron of twenty-nine B-24 bombers. Together they dropped more than seven hundred bombs. The raid was supposed to be a surgical strike with precision bombing. In fact, none of the bombs scored a direct hit.[5] The company's power plant was damaged, but the high concentration facility where the heavy water was produced was untouched. It was the locals who paid the steepest price: twenty-one Norwegian civilians were killed.[6] The Allies didn't know it at the time, but that last raid convinced the Germans that it was futile to keep repairing the plant. Instead, they decided to dismantle the whole operation and send it back home.

Heisenberg's quest to build a nuclear reactor suffered one other blow in the fall of 1943. At a secret three-day conference in Berlin, forty-four Uranium

Club scientists read papers summarizing their research results. One researcher had examined whether the uranium plates that Heisenberg favored were the best possible shape for a nuclear reactor and concluded that they were the *least* favorable design. He reported that Diebner's design casting uranium into an arrangement of small cubes was far superior. The only one in favor of using plates was Heisenberg. Karl Wirtz, the lead member of Heisenberg's reactor design team, appealed to Heisenberg to adopt Diebner's plan, but to no avail. Heisenberg could not accept that the reviled Diebner had come up with a better idea.[7] He was sticking with his layer design. His reason: "because theory was so much easier with simple layers of plates than with complex lattices of cubes."[8] When it came to "common-sense engineering physics[,] . . . Heisenberg was simply not very good," writes historian Jeremy Bernstein.[9]

CHAPTER THIRTY-THREE

"NICHOLAS BAKER" AT LOS ALAMOS

General Groves arrived at Los Alamos at the end of December 1943 with a special guest who was traveling under the name Nicholas Baker. Oppenheimer immediately recognized the distinguished Danish physicist Niels Bohr, the man who had identified U235 as the source of atomic energy.

Bohr, whose mother was Jewish, had returned to his home in Copenhagen after his stint at Princeton University. When the Nazis invaded in 1940, they chose not to roil the waters by arresting Danish Jews. But that changed in 1943, and Bohr, with the help of British intelligence, escaped to England.

Bohr had had a near twenty-year father-son relationship with Heisenberg. They first met in the summer of 1922 when Bohr was a guest lecturer at Göttingen and Heisenberg one of his students. Bohr took a shine to the brilliant young German, and in the decade that followed, the period when Heisenberg produced his greatest work, Bohr was his collaborator—but also his chief critic.

The two had vacationed together at Bohr's ski cottage and gone on long sailing trips together. Heisenberg had played with Bohr's children, taking them on pony rides.[1] But the war severed Heisenberg's relationships with many friends on the Allied side. None of the lost friendships pained him more than that of Niels Bohr. Two weeks after the war began, Heisenberg wrote his mentor an affectionate farewell: "Since I don't know whether and when our destiny will lead us together again, I will once again thank you for . . . everything you have done for me."[2]

Two years later, in September 1941, Heisenberg scheduled a lecture in Nazi-occupied Copenhagen and used the occasion to reach out to his old friend. Bohr reluctantly agreed to see him, inviting his former protégé to dinner at his home. The get-together began on a sour note when Bohr asked Heisenberg about

Niels Bohr and Werner Heisenberg. PHOTOGRAPH BY PAUL EHRENFEST JR., COURTESY AIP EMILIO SEGRÈ VISUAL ARCHIVES, WEISSKOPF COLLECTION.

a comment he'd reportedly made at a recent luncheon defending Germany's destruction of Poland. Heisenberg responded that Poland was a tragic case but tried to excuse his countrymen, pointing out that Germany had not been as brutal in France.[3]

Heisenberg's true intentions in coming to Copenhagen would be the subject of much debate after the war. Heisenberg claimed that he came to warn Bohr that Germany was working on a bomb and figure out a way for scientists to join forces to halt its development. Bohr believed that Heisenberg's intentions were anything but humanitarian and that he had come solely to pump him for information about the Allies' bomb program.[4] When General Groves solicited Oppenheimer's take on the meeting, Oppenheimer agreed with Bohr.[5]

During their dinner in Copenhagen, Heisenberg sketched a cylinder with a line sticking out to illustrate what he was working on. Bohr thought it was a bomb and brought the drawing with him to Los Alamos. Oppenheimer and colleagues saw instead a nuclear reactor and control rods.[6] Either way, Bohr's account of his face-to-face meeting with Heisenberg provided the most compelling evidence yet that the Germans were, at the very least, building a nuclear reactor and on a path that could lead to a plutonium bomb.

After Bohr left Los Alamos, General Groves pulled Oppenheimer aside to give him the substance of a recent intelligence report. Groves cautioned that the source might not be trustworthy, but according to him, Germany was *not* building an atom bomb. The two men looked at each other. Oppenheimer shrugged and said nothing, but his message was clear: it didn't matter whether the report was true, he and Groves would finish what they'd started.[7]

CHAPTER THIRTY-FOUR

GLENN SEABORG

As 1943 ended, the Norsk Hydro heavy water plant was once again out of commission. The little intelligence General Groves had gathered on Heisenberg and the German bomb had pretty much dried up. Not knowing how far the Germans had gotten, Groves continued pressing his scientists and engineers to move faster.

A plutonium bomb remained their best bet for beating Heisenberg, but Oppenheimer and his colleagues were still operating in unfamiliar territory. No one had ever irradiated uranium in a nuclear reactor to manufacture plutonium. And, assuming it worked, Manhattan Project scientists had no proven method for separating the plutonium from the irradiated uranium.

Glenn Seaborg, the U.C. Berkeley chemist who discovered plutonium, led the Met Lab scientists in the search for an answer. They and the Du Pont engineers who would build the plutonium separation plants had spent months exploring different possibilities.

The process Seaborg favored worked in the laboratory, but it was complicated, and he couldn't be certain it would work at a full industrial scale. With Groves's impossible deadlines, he couldn't afford to wait for the completion of a small pilot nuclear reactor being built at Oak Ridge, and he was terrified of making a mistake. "What if the Germans beat us to the bomb because we'd picked the wrong process?" he would later write. "The entire war might turn on the question."[1]

Before coming to Chicago and the Met Lab, Seaborg had enjoyed a relatively stress-free academic life, lecturing, correcting papers, conducting research, and taking the summers off. Nothing had prepared him for the pressure of such a high-stakes life and death assignment. Unable to sleep, he spent ten days in the hospital recuperating from nervous exhaustion. The stress didn't abate until Du Pont and the Met Lab scientists finally agreed on a way forward.

The plutonium the Met Lab had used for its initial tests was manufactured in an atom-smashing cyclotron at Washington University in St. Louis. After running twenty-four hours a day for a year, it produced an amount of plutonium roughly the size of a grain of salt.[2] Then, in early 1944, the Met Lab started getting plutonium from the Oak Ridge pilot reactor. Each reactor sample was a thousand times larger than those coming from the cyclotron. As the stores of plutonium mounted, the Met Lab's medical staff became concerned.

Plutonium was a new element and not yet fully understood. The staff knew that radiation from plutonium couldn't penetrate the skin. The scientists could safely hold it in their hands, but no one could say for sure what might happen if a radiated speck of dust was ingested or inhaled.[3] For answers, Met Lab doctors turned to medical histories from the 1920s and '30s of women who had painted watch dials with radium to make them glow in the dark. It was common practice back then to lick the brushes to sharpen the points, so the women inevitably swallowed some of the radium.

Their medical records revealed an alarming number of patients suffering from anemia, bone fractures, disease of the upper and lower jaw, and a higher-than-normal incidence of cancer-related deaths.[4] Based on the new findings, Seaborg ordered staff to immediately retrofit the labs with hoods to filter the air and to install linoleum floors that could be easily washed.

Later studies would reveal that plutonium was one of the most toxic substances known to man. A thousandth of a gram breathed in on a tiny speck of dust would kill in a matter of hours or days.[5]

CHAPTER THIRTY-FIVE

MANUFACTURING PROBLEMS

CBS NEWS, DECEMBER 26, 1943, WINSTON BURDETT REPORTING

In the mountainous central sector of the Fifth Army front in Italy, American doughboys celebrated Christmas by pushing doggedly ahead in pouring rain, under wicked fire from enemy machine guns and knocking the Germans off another peak in the outskirts of San Vittore . . . It was a bitter fight for a key objective.

As the war entered its sixth year, U.S. forces were advancing against the Germans in Italy, but the Manhattan Project's engineers were bogged down and struggling to harvest enough U235 for a bomb. The electromagnetic separation plant was down for repairs and morale among those directing the research on the second plant, the gaseous diffusion separation plant, was plummeting. A delegation of British scientists who had done the early work on gaseous diffusion had been shown the plans for the Oak Ridge plant and were extremely critical.[1]

Groves's engineers had come a long way. They invented a technology to protect the miles of pipe from the corrosive uranium gas. They invented pumps with plastic seals that didn't need lubricating, eliminating the danger of uranium gas reacting with grease to cause an explosion. (After the war, that plastic would be sold under the brand name Teflon). And they were on the verge of solving the most critical design problem—a flexible tubular filter for separating the U235 and U238 atoms. But the prototypes, including the Adler and Norris design, were manufactured individually and, despite the best efforts of the nation's most brilliant engineers, mass producing them remained a problem.[2]

Without the filters, plans to harvest U235 atoms using gaseous diffusion were doomed. Delivery of enough U235 to fuel a bomb would be delayed for months. All the time, effort, and money that went into designing the U235 bomb would have been for nothing. With the clock ticking, Manhattan Project scientists and administrators considered abandoning the uranium bomb and focusing exclusively on plutonium.[3] But Groves would not hear of it. Redundancy, redundancy, redundancy. He had to have a backup in case the plutonium problems proved insurmountable. Kellex engineers had narrowed their filter development plans to the Adler–Norris and one other design. Free of any budgetary constraints, Groves decided to go full bore, building plants to manufacture both designs.

CHAPTER THIRTY-SIX

NORSK HYDRO SHIPS HEAVY WATER

Attacks on the Norwegian heavy water plant had already cost more than fifty British, American, and Norwegian lives when, in February 1944, a shortwave operator with the Norwegian resistance fired off an urgent message that would almost certainly involve more fatalities.

The plant was preparing to ship fifteen tons of heavy water to a lab in Germany. Thirty-nine barrels of the precious liquid would be loaded into two freight cars and transported by rail to a nearby port where the cars would be transferred to a rail-ferry for the next leg of the journey across Lake Tinnsjå and then on to Germany.[1]

The Norwegian heavy water shipments were the weakest link in Heisenberg's atomic supply line. The Germans knew it, and the Allies knew it. With such a large shipment at risk, the Germans took extraordinary precautions. The plant already had armed guards, anti-aircraft guns, land mines, and machine guns. The Germans added still more troops and sealed off every possible entry except the main gate. They brought in a special army detachment backed by a regiment of SS police to guard the railroad tracks and ride the freight cars transporting the heavy water to the port. A squadron of spotter aircraft was positioned at a nearby airfield to assist with aerial reconnaissance.[2] Several troops were assigned to the port to guard the ferry.

The Norwegian resistance fighters had no good options. The plant itself was too well guarded, as was the train. Their best hope was to sink the ferry. To minimize the number of fatalities, a plant manager working with the resistance scheduled the crossing for a Sunday morning when there would be fewer passengers.

Previously, the Gestapo had executed civilians as retaliation for the resistance killing German soldiers. In one instance, they had laid waste to an entire town: burning every house, sinking boats in the harbor, killing animals and sending the whole population to concentration camps.[3] It was a terrible price to pay.

A resistance shortwave operator wired London, questioning whether the operation was worth the reprisals. Within hours, London responded: "Case considered. Very urgent that heavy water be destroyed. Hope this can be done without too serious consequences."[4]

Norsk Hydro's transport engineer, who was not a member of the resistance but was aware of the plot, would almost certainly be a primary target for Gestapo retribution, so the resistance arranged an airtight alibi. On the Saturday before the attack, he was taken to the hospital where he underwent an emergency appendectomy.[5]

On a frigid Sunday morning, the Hydro, a rail ferry with fifty-three passengers and crew, started its voyage across Lake Tinnsjå. Below deck, hidden in the bilge near the bow, were nineteen pounds of plastic explosives attached to an alarm clock. The resistance saboteurs had set the timer to go off forty-five minutes after the scheduled departure when the ferry would be over the deepest part of the lake.[6]

The alarm clock went off as planned. The explosives ripped through the hull, and twenty-six passengers and crew members died in the freezing waters.[7]

CHAPTER THIRTY-SEVEN

OSS KIDNAP PLOT

While Norway grieved, Colonel Carl Eifler, the former head of a U.S. Army guerrilla unit battling the Japanese in Burma, began assembling a team of operatives for a special mission: kidnapping Werner Heisenberg. Such proposals had been kicking around since October 1942, when the United States passed up an opportunity to seize Heisenberg during a visit to neutral Switzerland. Sixteen months later, at the request of General Groves, "Wild" Bill Donavan, head of the OSS, precursor to the CIA, resurrected the idea.

Eifler, reputed to be "the toughest, deadliest hombre in the whole OSS," was given a budget of $100,000 ($1.3 million today, adjusted for inflation) and unprecedented authority.[1] OSS officials were notified that Eifler's mission had the highest priority and "must be made to work."[2] The first task would be to track down Heisenberg, who had moved his lab after it was bombed. In May 1944, General Groves and the OSS were finally confident they had located him in southern Germany.

By then, Eifler had worked up a preliminary plan to sneak into Germany from neutral Switzerland with a team of twelve operatives, kidnap Heisenberg, and drag him back across the Swiss border. From there, Eifler would arrange for a small U.S. Air Force plane to meet them in a remote Swiss valley, pick them up, and fly them to Great Britain. That plan fell apart when Eifler's superiors informed him that landing a U.S. military plane in Switzerland would be a violation of its neutrality. If discovered, there could be serious diplomatic problems. But it was the final destination, Great Britain, that presented the most serious issue. The OSS and the British Secret Intelligence Service (SIS) were at war with each other. SIS did not approve of the OSS running operations in northern Europe, so the whole kidnap plot had to be kept secret from them.

Forced to amend his plan, Eifler came up with a plot worthy of a James Bond spy novel. He and his squad would sneak into Germany, kidnap Heisenberg, and take him back across the Swiss border. There, Eifler and Heisenberg would board a small plane and fly to the Mediterranean where they would parachute into the sea and be picked up by an American submarine.[3] And if that didn't work? If the Swiss police arrested him, what would he do then? Kill Heisenberg and suffer the consequences.

Eifler spent months preparing. An aide to General Groves provided the operational intelligence: a photo of Heisenberg, his location, his daily routine, plus all the information necessary to plan a clandestine entry into Germany. Meanwhile, Eifler recruited his crew of a dozen men. The first two were men who had fought with him in Burma. The rest were men he had worked with before the war as an undercover U.S. Customs intelligence officer on the Mexican border.[4] They had all signed on knowing it would be a dangerous mission, but none knew the exact details.

The mission was so secret that not even career OSS officers were allowed to know what Eifler was up to. To maintain security, he spent weeks building a plausible cover story, claiming to be head of a strategic trial unit schooling OSS stations in the use of technical gadgets that the OSS research and development had invented for clandestine warfare. While his men trained at the OSS Field Experimental Unit in Washington, Eifler traveled from Britain to Italy, then to North Africa and India, demonstrating exploding pencils, time detonators, flashlight batteries with secret compartments for messages, a silenced .22 caliber pistol ideal for assassinations, and more.[5]

He was so convincing that British agents showed off some of their "toys and gadgets" like suppositories for smuggling microfilm and shared information about British clandestine tradecraft. Eifler passed those secrets on to American military officers in New Delhi, and when British intelligence heard about it, they were furious. The transgression caused General Groves and General Donavan to lose confidence in Eifler. As a result, the whole mission was scrubbed.[6] The aborted kidnap plot was the Allies' second attempt to remove Heisenberg from the race. It wouldn't be the last.

CHAPTER THIRTY-EIGHT

UNDERGROUND BUNKER

In Germany, there were only two companies—Degussa and Auer—capable of manufacturing the metal uranium plates that Heisenberg and Diebner needed for their experiments. After Allied bombs leveled the Degussa plant, there was only one which triggered a competition between Heisenberg and Diebner over whose order would be filled first. Not surprisingly, Heisenberg won. But the Nazi regime still saw building nuclear reactors as a low priority, so Auer was not able to deliver the large uranium plates until January 1944.[1]

After receiving the plates, Heisenberg and his head of construction Karl Wirtz got to work building and testing their next generation reactor, the *B-VI*, in their new underground bunker in Berlin. The bunker had two-meter-thick walls of iron-reinforced concrete. Inside, the main laboratory had a pit for the cylindrical reactor and the water that would surround it. It had air and water pumps which could be activated in the event of an accident, a remote-control device for handling radioactive materials, heavy water tanks with a purification system, and separate spaces for a workshop and living quarters.[2]

The bunker was superbly equipped, but the nightly bombing raids and frequent interruptions in water, electricity, and other supplies still made work difficult. By the end of 1944, Heisenberg and Professor Wirtz had run four trials with the *B-VI* reactor, which produced an encouraging 206 percent neutron multiplication. The next prototype, the *B-VII* reactor, would incorporate one important change: instead of using water as the main reflector to bounce neutrons back into the reactor, it would be wrapped in graphite. Experiment *B-VII* marked the first time Heisenberg had used graphite. It pushed the neutron multiplication rate even higher, but still not enough to sustain a chain reaction.[3]

The *B-VII* experiments were conducted with a significantly smaller staff than previous experiments. After Minister Speer directed the Uranium Club to find locations outside Berlin away from the Allied bombing, fifty-five members of the Kaiser Wilhelm Institute staff relocated to the town of Hechingen in southern Germany.

Walther Gerlach, the Uranium Club's Nazi supervisor, had taught in a nearby town. He knew the area well and had recommended Hechingen, a peaceful little village which had an old textile mill that could serve as workspaces. Gerlach also found a safe location for Heisenberg's test reactors in Haigerloch, a picturesque nearby town with cobblestone streets and buildings dating back to the 1100s.[4] The two towns had the advantage of being closer to Heisenberg's family in Urfeld than Berlin and farther from the advancing Soviet troops.[5]

Heisenberg would not move permanently to Hechingen until January 1945, but his relocated staff were soon joined in the area by Otto Hahn, the man credited with discovering nuclear fission, and his staff who had fled Berlin after their labs were damaged in an air raid. Carl Friedrich von Weiszäcker, Heisenberg's young protégé, and his family also took up residence in Hechingen after the Allies captured his previous base in Strasbourg.

Heisenberg had seen little of his own family in the past couple of years. Commuting between Leipzig, Berlin, and now Hechingen, and traveling to occupied countries to give lectures left little time for anything else. Elisabeth had a difficult life in Urfeld. She and the children had suffered bouts of scarlet fever. The cabin's roof had partially collapsed under the weight of winter snow, and Heisenberg's absence didn't help.[6] Feeding her young family was always a struggle, and the property's rocky mountainous slope made growing vegetables impossible. In her letters to her husband, Elisabeth expressed feelings of incredible melancholy. She complained about the firewood and coal shortages and their drafty summer cottage, which was never warm enough in the winter. She also reminisced about her husband's music making and the time when they "breathed air smelling of good things and happily lived life to the full."[7]

Heisenberg had hoped to bring his family from Urfeld to the more convivial surroundings of Hechingen, but by the time he moved there permanently, all the available lodging had been taken, and he had to settle for a single bedroom in a friend's house.

CHAPTER THIRTY-NINE

PLUTONIUM PROBLEM

WINSTON CHURCHILL, "ON ALLIED VICTORIES," *BBC*, MARCH 26, 1944

We can now say, not only with hope, but with reason, that we shall reach the end of our journey in good order and that the tragedy which threatens the whole world and might have put out all its lights and left our children and dependents in darkness and bondage perhaps for centuries, that tragedy will not come to pass.

In late March 1944, Prime Minister Winston Churchill went on the radio to address the British public with a "a short talk on how we are getting on." Oppenheimer and General Groves were heartened by Churchill's optimism, but they couldn't rest easy as long as Heisenberg's whereabouts and the status of the German bomb were unknown. With progress on the U235 bomb stalled, they looked to the plutonium bomb as the best hope to counter the German threat. But Manhattan Project scientists had only a tiny amount of plutonium that had come from a Washington University cyclotron and several more preliminary experiments to conduct.

On April 5, 1944, Los Alamos finally received its first gram of plutonium from a pilot reactor in Oak Ridge. Nuclear reactors could manufacture plutonium faster and in quantities billions of times greater than the university cyclotrons. The Oak Ridge shipment meant that Los Alamos had entered the last lap in its race to build an atom bomb. They could now begin the final experiments to determine the critical mass of plutonium, but they had their work cut out for them. According to an official history of the Los Alamos Project: "At that time [the spring and summer of 1944] there was not a single experimental result that gave good reason to believe that a plutonium bomb could be made at all."[1]

Emilio Segrè, an Italian physicist who had studied under Fermi in Rome, would lead the critical mass investigation. He had been on his way to the United States to do research when Mussolini passed the law barring Jews from working in Italian universities. Segrè, who was Jewish, stayed in the United States and eventually went to work at U.C. Berkeley where he along with a team of other scientists discovered Plutonium. Segrè knew its eccentricities better than anyone. Depending on the circumstances, it could be as hard as glass or soft as plastic; it could also shrink when heated.

But as knowledgeable as Segrè was, he still had much to learn. After conducting some initial tests, he was shocked to find that the reactor-manufactured plutonium (Pu240) had one more neutron in its nucleus than the previous (Pu239) plutonium samples manufactured in a cyclotron. And that meant trouble, serious trouble, that could potentially eliminate plutonium as an energy source.

Segrè alerted Oppenheimer, and in Chicago, he and Fermi spent the next several months repeating the plutonium experiments, hoping that Segrè's findings would be proven wrong. But they weren't. Segrè's discovery meant that the detonator they planned to use for the Plutonium bomb—the gun-type detonator—wouldn't work.

The Pu240 reactor-produced plutonium "bullet" would split spontaneously *before* it entered the Pu240 cylinder. As a result, the explosion would begin prematurely and fizzle out.[2] Oppenheimer needed a faster detonator for the plutonium bomb, and there was none available.

The Manhattan Project had invested more than six billion in today's dollars building Hanford.[3] More than 130,000 workers had spent 126,000,000 man-hours constructing 540 buildings, 600 miles of roads, and 158 miles of railroad track. The plutonium bomb had been a colossal gamble. And now it seemed a colossal failure.[4]

General Groves knew he'd be held responsible, and, as he said in his autobiography, he had a good idea what the punishment would be. "I and all of the principal Army officers of the project . . . will spend the rest of our lives so far back in a Fort Leavenworth dungeon that they'd have to pipe sunlight in to us."[5]

Was this hyperbole? Of course. But the general had reason to be despondent. Given the painfully slow rate at which Oak Ridge was enriching U235 and the plutonium detonator problem, the general figured he would not be able to deliver a bomb until the end of 1945.[6]

CHAPTER FORTY

A SECRET MEETING

The meeting took place in the darkened balcony of the old Warner Theater in Washington, D.C. Philip Abelson, the U.C. Berkeley physicist whose research helped lead to the discovery of Plutonium, was working for the U.S. Navy in Philadelphia and had gotten wind of the Manhattan Project and the problems at the U235 separation plants. He believed he could be helpful, but he knew that the navy meddling in a top-secret, army-run operation would not be welcomed, so he arranged the secret meeting with an employee of the Bureau of Ships in the hopes of opening a back channel to Los Alamos and Oppenheimer.[1]

Since leaving U.C. Berkeley, Abelson had been exploring ways to use nuclear power to propel a submarine. And he, like everyone else, had run headlong into the problem of separating and harvesting those energy-rich U235 atoms. But unlike everyone else, he elected to use a separation technology called thermal diffusion and in November 1943, convinced the navy to build a pilot plant at the Naval Boiler and Turbine Laboratory in Philadelphia. The Boiler lab generated copious amounts of steam. Thermal diffusion required a steady source of steam, so it was an obvious choice.

Abelson's plant contained a row of one hundred forty-eight-foot-high pipes. Inside each of the pipes was a cold pipe, and inside that a narrower steam-heated pipe. When a liquid form of uranium was pumped into the space between the walls of the hot and cold pipes, the heavier U238 atoms gravitated toward the wall of the cold pipe and eventually sank to the bottom where they were collected. The lighter U235 atoms flowed toward the hot wall and floated to the top of the pipe where they were also collected.[2]

Following the clandestine meeting, Abelson's contact got in touch with a friend in the Navy Bureau of Ordnance who passed his message to Edward Teller at Los Alamos who passed it along to Oppenheimer.[3] The timing could not have been more opportune. The gaseous diffusion separation plant at Oak Ridge was still sidelined, awaiting a solution to the tube/filter design problem, and the electromagnetic plant, once thought to be the best bet for separating U235, was barely functional.

Prior to Abelson's overtures, Oppenheimer had looked upon the two Oak Ridge separation plants as competitors in a high-stakes, winner-take-all horse race. After Abelson contacted him, he realized that the horse race analogy was a "terrible scientific blunder."[4] Instead of competing, the horses should work together as in a relay race. Abelson's horse—thermal diffusion—was fast out of the starting gate and would run the first leg. It would manufacture slightly enriched U235, which it would pass to electromagnet-driven Calutron. More of a long-distance, endurance runner, Calutron would take the slightly enriched uranium and keep processing it until it reached a bomb-grade 80 to 90 percent U235 concentration. The two horses/plants working together would shorten the processing time for bomb-grade U235. And, when and if gaseous diffusion became available, the time should improve even more.[5]

If Oppenheimer had told Groves how he learned about the navy plant, the general would have been irate because of the security breach. Instead, Oppenheimer concocted a story about a Los Alamos staff member meeting Abelson by chance on a visit to the navy yard. Groves immediately signed on to Oppenheimer's idea and went to work.

The prototype plant that the navy had built was far too small to meet the Manhattan Project's needs, and there was no time to custom design one specifically for Oak Ridge. So Groves searched for a company that would build twenty-one replicas of the navy's plant and string them together side by side in a giant 522-foot-long chain of plants with more than two thousand thermal diffusion pipes. To make the job even more challenging, he demanded that the construction be completed in just four months.[6]

Twenty-one companies turned down the job before Groves finally found two foolhardy enough to take on the challenge. Groves then reconsidered his four-month schedule and demanded that the plant be operational in just ninety days. To save time, the companies began training the workers who would operate the plant at the navy lab in Philadelphia while constructing the plant in Oak Ridge. By now, innovative shortcuts and impossible deadlines had become standard practice throughout the Manhattan Project. This time, unfortunately, the press for speed turned deadly. An explosion at the navy lab killed two workers and injured several others. Prudence dictated that the companies postpone operations until the cause could be determined and the problem corrected. Predictably, Groves refused to permit any delay. The contractor ended up completing the job in a miraculous sixty-nine days from the time they got the production order.[7]

CHAPTER FORTY-ONE

WUNDERWAFFEN

> **"RECOUNTING THE LANDING," CANADIAN RADIO, JUNE 8, 1944**
>
> Five minutes will always be vivid in my mind: shells were falling on the beaches, mines still exploding, the whole beach covered with small crafts, men at work organizing the beaches already, bulldozers widening the exits . . . ammunition and tanks and supplies and vehicles and guns coming ashore as far as the eye could see.

The long-awaited Allied invasion of German-occupied Europe had finally begun. As Allied troops fought their way inland from Normandy's beaches, General Groves monitored the war news looking for signs of radiation poisoning. Fortunately, there were none. But a week after D-Day, early in the morning of June 13, 1944, a flying bomb, flames shooting from its tail pipes and a loud buzzing noise coming from its engine, appeared in the night sky over the English Channel. Once inland, it flew low over the Thames River at a speed of four hundred miles an hour. The rocket appeared headed for London, but if London was the target, it never made it. A short distance from the town of Swanscombe, the engine cut out, the flames extinguished, and the buzz went silent. Cloaked in darkness, the rocket disappeared for several terrifying moments until 4:18 a.m. when a flash of light and loud explosion marked its final location.[1] The rocket hit a railway bridge, demolishing the bridge, killing eight civilians, and wounding many more. Enraged and humiliated by the Allies' successful invasion, and eager for revenge, Hitler began deploying one of his top-secret vengeance weapons: the V-1 flying bomb, the latest entry in the "Secret War" pitting German weapons engineers against Allied engineers.

British authorities had been expecting this moment and were attacking V-1 launch sites ever since the RAF raid on Peenemünde ten months earlier. With

Allied forces pushing inland and headed for Berlin, Hitler was making British civilians pay a price, except this time there would be no German pilots lost. The V-1s were not accurate. Once launched in the direction of a target, they could not be guided, so they served little strategic purpose. But the loud buzzing noise signaling their arrival, and the terrifying moments of silence during which everyone within earshot became potential victims, made them a very effective terror weapon.

Four more V-1s struck that night. The V-1s exacted a terrible toll in physical damage and lives lost, but it was the unseen threat and potential for even greater suffering that worried British officials the most. The Germans were reportedly using poison gas in their death camps to murder their prisoners. It was not unthinkable, therefore, that Adolf Hitler would use radiation poisoning to gain a battlefield advantage.

By coincidence, the Alsos scientific intelligence unit was in London preparing for its mission when those first V-1s struck. British intelligence contacted its leader, Samuel Goudsmit, and asked him to check for radiation.

The next day, dressed in protective clothing and carrying a handheld radiation detector, Goudsmit descended into the crater left by one of the V-1 explosions. While MI6 officers watched from above, he picked his way slowly through the debris, making certain he covered the whole crater. The needle on the detector barely budged. The bomb's one-ton warhead was not radioactive. The German Wehrmacht would fire more than 9,500 V-1 flying bombs at Great Britain in the coming months. None of them spread radioactive debris. But they did confirm Hitler's penchant for fantastical, out-of-the-box weapons. His engineers had introduced navigation by radio beams and pilotless bombs. Soon to come: the V-2, the first long-range guided missile, and the jet plane—and, if Allied fears proved correct, an atom bomb.

CHAPTER FORTY-TWO

DARKEST HOURS

CBS NEWS, AUGUST 6, 1944, BILL DOWNS REPORTING

Senior British staff officers have said tonight that the German Seventh Army assigned to hold the Allied troops within their Normandy beach head have suffered a major defeat and that the events of the past ten days have given the American, British, and Canadian forces in France their first big victory.

As Allied troops advanced in Europe, Heisenberg continued testing his *B-VII* experimental reactors in the Berlin bunker. Meanwhile, Los Alamos, with both bombs dead in the water, entered what would become its darkest hours.

Twenty-two months after the launch of "the greatest scientific gamble in history," Robert Oppenheimer and General Groves faced the very real possibility that they would fail. Los Alamos would lose the race for the bomb. Nazi rule with all its horrors would spread across Europe, the Soviet Union, and beyond. Armed with an atom bomb, Hitler could reverse whatever gains the Allies might make. Isolated atop a mesa in the New Mexico desert thousands of miles from the fighting, Oppenheimer and crew felt as if the entire weight of the Allied cause rested on their shoulders. "The only way we could lose the war," one concluded, "was if we failed our jobs."[1]

The possibility of failure weighed heavily on Oppenheimer, who became uncharacteristically irritable and short tempered. Physicist Isidor Rabi recalled the sense of gloom that pervaded a meeting at the height of the crisis. "Who were the German scientists? We knew them all. What were they doing? . . . We finally arrived at the conclusion that they could be exactly up to us, or perhaps further. We felt very solemn."[2]

D-Day had come and gone. Allied armies had begun a long, bloody march to Berlin. Groves's worries that the Germans would use radioactive contaminants to block the Allies' advance had proved unfounded, but the fear of a German atom bomb persisted.

Desperate, Oppenheimer reorganized Los Alamos around one single mission: developing a faster detonator for plutonium. Though unaccustomed to working together, scientists and engineers, electronics and explosives experts, and chemists and metallurgists all pooled their knowledge and expertise. Delay meant the deaths of more Allied servicemen and civilians and ratcheted up fear that Heisenberg would build the bomb first, that Hitler would target London, Moscow, and perhaps New York.

The quest for the bomb took Manhattan Project scientists, engineers, and their corporate partners down roads never before traveled. And each presented its own unique obstacles demanding new methods and new technology to propel the project forward. After the war, Pentagon officials would estimate they had produced more than five thousand patentable inventions.[3] No obstacle proved more intractable than the design of a detonator for the plutonium bomb.

The issue of detonators had been addressed generally in an ordnance meeting at Los Alamos in April 1943. At that gathering, Seth Neddermeyer, a Caltech protégé of Oppenheimer, suggested using an implosion device. Oppenheimer initially rejected the idea but then changed his mind and gave Neddermeyer the green light to work on it.[4] At the time, the notion seemed far-fetched. Now, it was Oppenheimer's only option.

The fraction of a second it would take the gun-type detonator to assemble the critical mass was simply too slow for a plutonium bomb. The implosion detonator would be faster but infinitely more complex: a high-performance race car versus a Model T. Basically speaking, it would mount thirty-two high-energy explosive charges symmetrically and equidistant from each other on the outer shell of the spherical bomb. The detonators would ignite their fast-burning explosives at precisely the same moment, channeling their detonation waves inward. Inside the bomb, the leading edge of each convex wave would run into a cone-shaped slow-burning explosive, which would delay the front of the wave, allowing its outer edges to catch up and move ahead. As a result, the convex waves would be transformed into concave waves. The concave-shaped waves would then wrap around a subcritical sphere of plutonium with a four-centimeter radius and compress it into a three-centimeter sphere. The smaller, more dense mass would cut the distance neutrons had to travel to split another nucleus. At the same time, an initiator buried inside the critical mass would spew out millions of neutrons, accelerating the chain reaction and cutting the time from detonation to explosion to mere microseconds.[5]

Every step in the process had to be perfectly choreographed. If the shockwaves from the explosions overlapped, if they failed to cover the entire surface of the plutonium sphere, if one shockwave lagged the rest by even a millionth of a second, the critical mass would deform, neutrons would escape through the deformities, and the bomb would fizzle out.[6]

In the fall of 1944, with Los Alamos in crisis mode, Oppenheimer and General Groves scoured the country looking to recruit additional mathematicians and explosives experts to work on the implosion detonator. Groves arranged to have six hundred members of the army's Special Engineering Detachment reassigned to help with the exhaustive testing required.[7] He relaxed the tight security at Los Alamos so that British scientists, whom he had previously barred from the top-secret site, could be brought in, a move he later came to regret.[8]

Among the British scientists was a young German-born British citizen named Klaus Fuchs, a refugee from Nazi persecution whom Laura Fermi got to know after she and Enrico moved to Los Alamos. Laura was invited to a picnic at a site with well-preserved cave dwellings near the Los Alamos mesa. She was afraid to drive down from the mesa on the narrow dirt roads, so Fuchs volunteered to drive her. Laura described him as "extremely polite . . . and refined," but mostly quiet. British officials claimed that they had put him and the other British scientists through a careful security check. But even a cursory check should have discovered that Fuchs had been interned for seven months in a British camp for enemy aliens early in the war and might harbor some resentment toward the British. In 1950, he was convicted of spying for the Soviet Union and served nine years in prison.[9]

For months, the walls of Little Pajarito Canyon shook from hundreds of test explosions as some of the world's most brilliant explosives experts attempted to design an implosion detonator for the plutonium bomb. Watching from concrete bunkers and shielded behind bulletproof glass, they attempted to configure explosive charges around a large metal sphere so that they compressed it into a perfectly symmetrical smaller one. The tests were filmed by a remarkable new sweeping-image camera that had a shutter speed of one hundred millionth of a second.[10]

The results were invariably disappointing: One point on the sphere would sink in faster than another, or not at all. Instead of a perfect sphere, they would end up with one that was dented or lopsided.[11] It took six months, but after countless failed attempts, an army of experts finally appeared to be nearing their goal. If all went well, Oppenheimer's scientists estimated the plutonium bomb would produce an explosion even more powerful than that of the U235 bomb.[12]

But everything did not go well. Manufacturing the precision components proved more difficult than anticipated. The first shipment of parts failed at an alarming rate, and the contractor supplying the firing circuits was late on delivery.[13]

CHAPTER FORTY-THREE

ALSOS IN PARIS

> **"PARIS DELIVERED," FRISCH PATHÉ GAZETTE NEWSREEL, AUGUST 1944**
>
> As Allied troops sweep onwards toward Paris, a three-word announcement from the capital brings the great news: "Paris is liberated." Cheering crowds line the streets as the Allies speed through the outskirts.

While Parisians celebrated, Alsos, General Groves's new scientific intelligence unit, rolled into the City of Lights. Their mission: to locate Germany's atom bomb facilities, seize its uranium, and capture its atomic scientists before the Soviets could nab them. The mission's military commander was a man Oppenheimer knew all too well: Colonel Boris Pash, the fervent anti-Communist counter-intelligence officer who had hounded him so mercilessly at Los Alamos. His "thorough competence and great drive had made a lasting impression" on General Groves, who appointed him to head the mission.[1] True to his reputation, Pash was among the first Allied troops to enter Paris. Driving in an open jeep, he and his men weaved in and out of a column of French tanks dodging sniper fire as they went.

Goudsmit arrived in Paris a week after Pash. A nuclear physicist adept at problem solving, he was also well suited for intelligence work. While in college, he had studied Egyptology, learning how to decipher hieroglyphics, and had taken a forensics course taught by a local police chemist where he learned how to compare fingerprints and use ultraviolet light to detect forgeries. "Sam has a sixth sense when it comes to bringing order to jumbled facts," says I.I. Rabi, a fellow student at the University of Leiden. "He thinks like a detective. He *is* a detective."[2]

Goudsmit's gumshoe skills would be a boon to the mission, but it was his web of personal relationships with Heisenberg and other European physicists that made him truly indispensable. In a letter volunteering for the assignment, he had written, "I have very close personal contacts with most of the physicists in Italy, France, Belgium, Holland and even Germany. I think there are even some German physicists who still believe I am their friend"—a reference, no doubt, to Heisenberg.[3] Goudsmit mailed his letter on June 25, 1943, three months after receiving a letter from his parents whom he had last seen in Holland before the war. It bore a Portuguese postmark, but it had been written inside the barbed wire fences of Auschwitz, the notorious Nazi death camp.

Pash and Goudsmit stayed in Paris for three months, pursuing intelligence leads while waiting for Allied troops to drive the Germans out of France. Their first stop was the laboratory of Frédéric Joliot-Curie. He and his wife Irène were pioneers in nuclear fission research. More importantly, Frédéric had allowed German scientists to use his cyclotron. If anybody knew what the Germans were up to, Goudsmit figured it would be Joliot-Curie. But Joliot-Curie said he knew nothing about German atom bomb plans. He did suggest, however, that they try to find Kurt Diebner. He would know. It was the first time Goudsmit heard the Diebner name.[4]

Broadening their search, Goudsmit and Pash raided the deserted offices of German electronics and chemical firms. A discarded piece of carbon paper at one company gave them the names and addresses of French employees. Two months' worth of telephone messages as well as the sign-in book in the lobby yielded names, affiliations, times, and dates of visitors. Desperate for leads, Goudsmit and Pash tracked down several employees and visitors looking for clues to anything related to atomic energy.

Their detective work eventually led them to a Herr Petersen, the manager of the Paris office of Auer, a large German chemical company. Peterson had fled the city, but just before the liberation of Paris, he had sent a large shipment of the radioactive element thorium via rail to Auer offices in Berlin. "This really scared us," said Goudsmit, who believed it could be used to fuel an atom bomb. Thorium was also three times more abundant than uranium.[5]

Checking industrial uses for the chemical, Goudsmit discovered that very few companies used it and only in very small quantities. He found no reason for such a large shipment, except building an atom bomb. Goudsmit, by his own admission, became obsessed by the thorium mystery.[6]

Pressing further, he asked Auer's French employees for any company files that had been left behind. Among them he discovered a notebook listing outgoing registered letters. Almost all of them had gone to Auer offices in France or Germany, and all but the last one had a postage stamp. That exception was addressed to Fräulein Wessel (Petersen's secretary) in Eupen—a town on the border of Belgium and Germany. Because it had no stamp, Goudsmit assumed the letter had been hand delivered.[7]

Based on that sliver of evidence, Pash decided to drive to Eupen and search for Wessel. The French Army had driven the Germans out of the town and now administered that area. Allied protocols required that American soldiers get French authorization to operate there, but Pash and his team ignored the rules, climbed in their jeeps, and drove the 250 miles to Eupen. Once there they hit pay dirt. They found not only Fräulein Wessel, but also her boss Herr Petersen and a suitcase full of documents. Petersen would become the Alsos mission's first prisoner.

Back in Paris, Pash and Goudsmit spent several hours interrogating Petersen, but he claimed to know nothing about the thorium. Frustrated and disappointed, Goudsmit combed through the contents of Petersen's suitcase. He found a ticket from a Berlin streetcar and a recent bill for a hotel in Hechingen, a small village in southern Germany. The name rang a bell. After his laboratory in Leipzig was bombed, Heisenberg had eventually moved operations to southern Germany, and Allied intelligence believed that the small town of Hechingen might be his current location. Goudsmit was elated. "It seemed a tight case," Goudsmit would later write. It connected the thorium in France with the town of Hechingen and Auer, Germany's foremost chemical company.[8]

But Goudsmit's "tight case" soon sprung leaks. Petersen's mother lived in Hechingen, and he had gone there to visit her. The thorium? Looking to the company's future after the war, Auer planned to sell thoriated toothpaste for "sparkling brilliant teeth." They had sold it during the early '30s with advertising that featured armies of little radioactive men fending off evil bacteria.[9] An American company, makers of Pepsodent toothpaste had boosted sales with a whitening agent it called irium. *The Pepsodent Show Starring Bob Hope* advertised the benefits of irium every week on Armed Forces Radio. If irium worked for Pepsodent, why not thorium and its army of little radioactive men for Auer's toothpaste?[10]

Goudsmit's detective work did yield one potentially significant piece of intelligence. A letter from Petersen's mother referred to the town of Hechingen as a restricted area. Goudsmit passed the information along to General Groves and British intelligence.

While in Paris, Goudsmit also came across a course catalogue for the Reich University of Strasbourg. A model of German higher education and part of a propaganda program to spread fascism in France, it had been opened in 1941 and was staffed primarily by Nazi loyalists including some atomic physicists. Goudsmit and Pash were eager to get to Strasbourg and interrogate those men. Unfortunately, the city on the French–German border was still under German control, so they had to wait for it to be liberated.

The two used the time to take a side trip to two newly liberated cities. In Brussels, Goudsmit raided the offices of Union Minière, a large mining company that held most of the world's known uranium deposits. The company's books showed shipments of more than a thousand tons of uranium products to Germany since 1940, far more than German firms would need for normal commerce and a near certain sign they were attempting to harness atomic energy.[11]

From there, Goudsmit moved on to the Dutch city of Eindhoven, home of a plant suspected of supplying nuclear research equipment to the Nazis. It turned out that Goudsmit and a young physicist working at the plant had a friend in common. Goudsmit asked if that friend had any news about his parents. The young man confirmed the Germans had deported them along with other Jews.

Back in Paris, as he wandered the streets, Goudsmit could not stop thinking about his parents. When he was growing up in the Hague, his mother owned a stylish French hat shop named Au Louvre. He loved hearing her talk about designing hats, about the challenges of predicting next year's fashions, and, of course, about Paris, the world's fashion capital. The young Sam Goudsmit had hoped to go into the millinery business himself one day. "Even now, when I think of Parisian streets and buildings," he told a reporter, "I can hear my mother's voice telling me about them."[12]

CHAPTER FORTY-FOUR

HANFORD

In September 1944, while Goudsmit waited for Allied troops to liberate Strasbourg and Oppenheimer pushed his engineers and scientists to perfect the implosion detonator, Enrico and Laura Fermi prepared to leave Chicago and relocate to Los Alamos so Enrico could help with the detonator. They had sublet the house and hired movers, but then Enrico got an unexpected call: he was needed for a secret mission. While Laura and the children headed to New Mexico by train, Fermi departed on a different cross-country trip. His destination: Washington state.[1]

Obsessed as usual with security, General Groves required that Fermi travel with a bodyguard and use a false name, Henry Farmer. The pair boarded the next train for the Pacific Northwest and got off in Pasco, Washington, well after midnight.[2] A car drove them the rest of the way to the sprawling 560-square-mile Hanford Engineer works.

Later that morning, Du Pont's head engineer and a Met Lab associate of Fermi took him on a tour of Hanford. They passed Richland Village, a community of conventional wood-frame and prefabricated houses, quarters for engineers and administrators, and Hanford Camp where some fifty thousand blue-collar workers were billeted in barracks and mobile trailers.[3]

The tour continued about forty miles north through high desert scrub brush and tumbleweed to a large construction site where two enormous structures, eight hundred feet long by sixty-five feet wide by eighty feet high with six-foot-thick walls, were taking shape. There, engineers working via remote control would orchestrate the final steps involved in separating plutonium from the irradiated uranium.[4]

Fermi's tour ended on the south bank of the Columbia River where Nuclear Reactor B, one of Hanford's three nuclear reactors, was located.[5] Its design, like that of the remote-controlled separation plant, was yet another Manhattan Project first-of-a kind technology. "It had all kinds of tubes and pipes running through it. Nobody could understand what kind of a contraption it was," commented one pipefitter recruited to work on it. "They had never seen anything like it."[6]

Waiting inside that contraption was a small group of engineers and reactor operators. They greeted Fermi in front of a massive 1,200-ton graphite block that was thirty-six feet deep, twenty-eight feet wide, and thirty-six feet high, surrounded by walls of steel and concrete several feet thick. More than a thousand inch-wide channels drilled into the face of the graphite block were waiting to be filled with uranium fuel rods.[7] Everyone was nervous. The reactor was about to undergo its first full-scale test.

At 5:43 p.m. on Wednesday, September 13, 1944, Fermi began the delicate process of inserting the fuel rods. The loading continued over a day and a half with Fermi stopping periodically to check the neutron activity as he had done two years earlier in Chicago. On Friday morning, as the reactor began to approach criticality, the operating crew inserted neutron-absorbing control rods along with the fuel rods to slow the nuclear reaction. From that point, Fermi checked his slide rule after every fuel rod was inserted to make sure that the radioactivity was increasing as predicted. And the reactor began to heat up.

Finally, on September 27, shortly after midnight and two weeks after the loading began, Fermi ordered all the control rods to be withdrawn, triggering a sustained nuclear chain reaction. As it accelerated, the power increased. By 2:00 a.m., the reactor was generating a million times more power than Chicago's pioneering CP-1 reactor had.[8]

The mood in the building was euphoric. Fermi and colleagues had crossed the nuclear frontier. They had triggered a nuclear chain reaction at a power never achieved and had sustained it for longer than previously achieved.

But at 3:00 a.m., the reactor's engineers began whispering to one another and frantically adjusting the control rods. The power was dropping, slowly at first, then more rapidly. The chain reaction was dying. By 6:30 a.m., the reactor had shut down.[9] The normally unflappable Fermi was shaken. Dozens of scientists and engineers had worked to design the reactor, but as the lead scientist he felt responsible. His colleagues, and arguably the whole Manhattan Project, were relying on him.

Then, suddenly, mysteriously, the reactor began to stir. Slowly, gradually, the power picked up. And for several hours, it appeared that whatever had killed the chain reaction had cured itself. By 4:00 p.m. the next day, the reactor was back, generating nine thousand kilowatts, the same level where it had cut out.[10]

And then, once again, as if some malevolent force was toying with them, the chain reaction began to die. It seemed, despite the enormous strides of the past several years, the despairing scientists still had much to learn. There had to be an explanation. Sabotage? Leaks in the water pipes that cooled the reactor? No one knew. Briefed on what was happening, General Groves decided not to interfere.[11] For Oppenheimer, the reactor failure coming just months after the discovery that the plutonium detonator would not work felt like a one-two gut punch.

After consulting with scientists in Chicago and Oak Ridge, Fermi sent a messenger to fetch John Wheeler, a thirty-one-year-old Princeton-educated physicist. After leaving Princeton, Wheeler had been recruited to work at the Met Lab in Chicago and was now in Hanford where he focused on what he called "the bugs under the table"—all those things that could go wrong with a nuclear reaction.[12] When Wheeler arrived, he and Fermi went to work studying the data accumulated over the past few days.

A Manhattan Project colleague of Fermi, Richard Feynman, would later describe the study of physics as "something like a great chess game being played by the gods, and we are observers of the game. We do not know what the rules of the game are; all we are allowed to do is to watch the playing."[13] Fermi and Wheeler spent several hours "watching the playing," studying every move the "gods" made, and finally they understood.

Something was poisoning the chain reaction. An unknown substance was absorbing neutrons. That substance would build up over time, denying the chain reaction the neutrons needed to continue. The chain reaction would die, and the substance would eventually decay. Then, with the poison gone, the chain reaction would begin again. The pattern signaled that the poison had to be radioactive and short lived.

Fermi and Wheeler knew the effects lasted 9.4 hours, so they looked for a fission product whose half-life, its rate of decay, was also 9.4 hours and came up with Xenon-135. Armed with that information, Fermi calculated they would have to increase the number of uranium fuel rods by 25 percent to overpower the xenon. Fortunately, Du Pont's engineers had allowed a margin of error when they built the reactor. They had drilled more holes than the original plans called for, enabling them to accommodate the extra fuel rods. But rectifying the problem would cause a five-month delay—a delay they could ill afford.[14]

Meanwhile, Allied troops were attacking along Germany's western border, and an increasingly desperate Adolf Hitler threatened once again to unleash his wonder weapons. "Very soon I shall use my triumphal weapons, and then the war will end gloriously. . . . This is the weapon of the future, and with it, Germany's future is likewise assured."[15] In the beginning of October 1944, two news agencies, Reuters and the International News Service, quoted reports from Swedish scientists identifying one of those weapons and confirming the fears of Manhattan Project scientist: "Reports reaching neutral scientists . . . said the Germans already have made atom bombs experimentally in considerable numbers. Mysterious explosions seen 80 miles distant from Bornholm [Denmark] were caused by atom bombs."[16]

In another story, Reuters quoted an eyewitness who described the results of one of those tests: "The countryside for hundreds of yards around was completely flattened as if it was sheared by a knife."[17] Later that month, the OSS reported from Stockholm that it had received reliable reports that a German U-boat would sail to the eastern coast of the United States, where it would launch a V-1 flying bomb aimed at New York City. In a separate report, American military authorities warned that "the American east coast might be the area chosen for a blind attack by some sort of flying bomb. . . . This device is based on the principle of the explosion of the nuclei in heavy hydrogen derived from heavy water."[18]

STRASBOURG LIBERATED

French troops liberated the city of Strasbourg on November 23, 1944. Days later, Colonel Pash and an advance team of Alsos soldiers made a dash for the university where they took several German scientists into custody.[1] Goudsmit followed a week later.

With German artillery still shelling the city, he and his men spent two days and candle-lit nights poring over the papers of the captured scientists. In a letter to his wife, Goudsmit described the agony he felt knowing that he would be responsible for sending fellow scientists to prison. "I am too soft for this game," he admitted.[2]

Goudsmit was shocked by the information the prisoners' papers revealed about their research. One document listed members of the Uranium Club along with the addresses of their laboratories and institutes.[3] He also discovered an exchange of letters between C. F. Weizsäcker and Heisenberg discussing nuclear fission. One of those letters included Heisenberg's new address in southern Germany and his telephone number.[4] Goudsmit now had the location of most of the scientists on his most-wanted list.

Alsos investigators also found a ripped-up letter in a wastebasket drafted by one of the German scientists severely criticizing Heisenberg for some of his calculations. Goudsmit later discovered a revised copy of the letter among Heisenberg's papers. This version struck a more measured tone. The writer apparently had had second thoughts about so boldly criticizing Germany's pre-eminent nuclear physicist.[5] He wasn't the only one to question Heisenberg's calculations. One of his former students would later comment, "Though a brilliant theoretician, he was always very casual about numbers."[6]

Goudsmit and his Alsos colleagues had come to Strasbourg looking for research on nuclear fission and an atom bomb. What they found turned their stomachs. They saw firsthand evidence of Nazis atrocities, a small one-shower gas chamber in one of the laboratories, records of human experiments Nazi doctors had conducted on Soviet prisoners of war, and experiments conducted by Eugene von Haagen, a Nazi biologist.[7]

Von Haagen developed vaccines for influenza, yellow fever, and other diseases. His laboratory was filled with cages for mice, guinea pigs, and monkeys.[8] But he also had a second lab, this one located in a nearby prison, where he tested a vaccine for spotted fever. There they found even more gruesome evidence of Nazi depravity: large tanks containing human corpses floating in alcohol.[9] Von Haagen had fled Strasbourg, but a copy of a letter to a Nazi SS officer revealed the nature of that research. "Of the one hundred prisoners you sent me, eighteen died in transport," von Haagen complained. "Only twelve are in condition suitable for my experiments." He then requested another hundred before closing with "Heil Hitler."[10]

By hideous coincidence, Goudsmit was billeted in the von Haagens' apartment and slept in his son's bed. "All his toys were still there, his electric train, a movie projector, an old microscope," along with mementos of his time as a Hitler Youth, Goudsmit wrote his wife.[11]

In liberated Paris, Goudsmit managed to occasionally escape the stress of the war by taking long walks through the quiet streets. In Strasbourg, the horrors were present night and day. In the mornings, as he made his way from the von Haagens' apartment to the Alsos office, he had to thread his way past weary soldiers trembling from shell shock.

It was all too much for the genial college professor. The stress of the war, and the waning hope that his parents might somehow have survived, left him emotionally drained. Back at the apartment, he became inconsolable. He "just went off his rocker . . . weeping and thrashing about," an Alsos friend reported. It took thirty minutes to calm him down.[12] A short while later, Goudsmit flew back to America for a brief, much-needed rest.

The documents that Goudsmit discovered in Strasbourg were an intelligence gold mine. For the first time, the Alsos mission had a window into the inner workings of Germany's atomic research program: notes from secret meetings, postcards, memos from one scientist to another, and letterheads with addresses and telephone numbers.[13] Goudsmit was shocked that the Germans made so little effort to censor their communications.

It took several days for Goudsmit and colleagues to sift through all the documents. In the end they found no evidence of a massive industrial facility like Oak Ridge or Hanford, but they did confirm that Heisenberg headed a team of scientists working to build a nuclear reactor. Other teams were working to separate U235 from uranium ore and to produce heavy water. The cache of papers also revealed that the German Army, independent of Heisenberg, was building its own nuclear reactor, and the scientist in charge was Kurt Diebner, a loyal Nazi, but a complete unknown within the circle of world-class theoretical physicists.[14]

After reviewing all the documents, Goudsmit felt confident that the Germans were far behind the United States in the race for the atom bomb. The news reports that the Germans had already developed the atom bomb were most likely mistaken or Nazi propaganda. Goudsmit arranged to have his findings hand delivered to General Groves in Washington.[15]

But Groves's response was not what Goudsmit had expected. The general remained skeptical. Locating Heisenberg had been too easy; the documents with addresses and telephone numbers, too revealing. General Groves worried that they had been planted for the Allies to find, that they were Nazi disinformation.

It turned out that General Groves was right to be skeptical. Goudsmit had missed several critical pieces of intelligence. Among them, a document discussing a large-scale fission experiment which had taken place several miles outside Berlin and a memorandum stating that Hitler had been briefed on atomic weapons.[16]

The discovery of that memorandum coincided with a comment Hitler had made in a recent address to members of the military: "Up to this day I can answer for all the acts I've committed before God and my compatriots. But for what I'm going to order in the near future, I will no longer be able to justify before God."[17] This, from a man who had already murdered millions of Jews. What could possibly be worse?

CHAPTER FORTY-SIX

RESTRICTED AREA

By November 1944, the Alsos team had moved on from their thorium toothpaste adventure. General Groves had not. The Alsos investigation had left some critical questions unanswered: Why was the town of Hechingen a restricted area? What were the Nazis hiding?

The RAF had begun flying reconnaissance flights over hundreds of square miles near Hechingen looking for answers.[1] Their specially equipped Spitfires carried two cameras, one on each wing, which filmed overlapping images. Back in England, analysts from Britain's Air Scientific Unit studied the photos through stereoscopes that married the two images, producing what looked like a three-dimensional picture.

The interpreters spent countless hours looking for uranium processing plants and nuclear reactors. They had been briefed to search for a large industrial facility using enormous amounts of electricity. With so much territory to cover, the interpreters used Germany's electric power grid as a guide. They looked for large electric transformer stations and followed the high-voltage power lines radiating from them. It was painstaking, mind-numbing work, but what it turned up left General Groves genuinely scared.[2]

In a valley not far from Hechingen, they discovered a twenty-mile stretch with more than a dozen medium-sized industrial plants, all under construction. Each had a small factory building with storage tanks and what looked like a grid of pipes alongside them.[3] The interpreters observed workers from nearby slave labor camps building rail spurs and stringing power lines. They saw large amounts of material being offloaded from trucks and railcars. Every reconnaissance mission brought new evidence of what was obviously an urgent, top-priority program. But it was a second bit of intelligence that bumped the program to the top of General Groves's watch list: Werner Heisenberg had been spotted in the area.

As the interpreters grew more familiar with the imagery, they noticed that the dozen or so plants were all strung along the same geologic contour. Armed with that additional bit of information, the Interpretation Unit sent an officer to Britain's Geologic Museum. There, a curator searched through volumes of old records and discovered that German geologists had explored the area before the war. But they were not looking for uranium; they were looking for oil-bearing shale.[4]

Germany was desperately low on gasoline. Allied intelligence concluded that the pipes alongside the plants were used to "cook the oil out of the shale." What the reconnaissance had discovered was a crash program to refill Germany's gas tanks.[5] Allied intelligence later learned that the German term "restricted area," which they interpreted to mean "top secret, keep out," didn't mean that at all. It was simply a warning that the town, already flooded with people from war-ravaged areas, was no longer open to refugees.

CHAPTER FORTY-SEVEN

BATTLE OF THE BULGE

Goudsmit, fresh from his visit home, was back in Paris with Pash waiting for Allied troops to occupy southern Germany so they could get to Hechingen and to Heisenberg's lab. Meanwhile, 185 miles to the northwest in the Ardennes Forest, British and American troops were relaxing, looking forward to a peaceful Christmas holiday. The Germans were on the run. It seemed only a matter of time before the Allies would roll into Berlin, and the war would be over. But then at dawn on December 16, 1944, the roar of German artillery and the clanging of German tank treads shattered the calm. Bad weather had grounded Allied reconnaissance flights, and the Allies were caught completely by surprise. Twenty-five German divisions launched a ferocious attack.

"THE ENEMY STRIKES," THE WAR DEPARTMENT PRESENTS

In one day, the enemy smashed though the defenses of the American First army on a forty-five-mile front and was fighting deep into Luxembourg and Belgium.

Marshaling what was left of the once-mighty German Wehrmacht, Hitler had launched a brilliant, but desperate, attempt to encircle and crush U.S. and British forces and seize their fuel supplies. A German radio station announced that their forces were using a new type of bomb powered by atomic energy. "Wherever such a missile is dropped," the broadcast reported, "all animals and plants cease to exist. Huge areas of land are scorched, woods are consumed, and any human being caught in the hurricane is shattered to smithereens."[1] The announcement, picked up by the

Associated Press and widely reported in newspapers, turned out to be more German propaganda, an apparent effort to panic the Belgians.

The Germans' surprise attack nearly worked. But the skies cleared, and Allied air superiority and reinforcements eventually carried the day. The Allies had let their guard down and paid an enormous price. The Battle of the Bulge, as it was later called, cost nineteen thousand American lives. It was the bloodiest battle of the war for America's troops.

Heisenberg was in Zurich preparing to give a lecture and doing some Christmas shopping when he heard the news of the attack. Paul Scherrer, a physicist at the city's Federal Technical College, a friend of Goudsmit, and, unbeknownst to Heisenberg, an important intelligence source for the Allies, had invited him to speak. Scherrer cabled Goudsmit to tell him Heisenberg was coming. Goudsmit then notified the OSS, which assigned one of its top agents, Moe Berg, to attend the Zurich lecture. A man with a colorful history, Berg had studied seven languages at Princeton and played fifteen years of professional baseball, ending his career as catcher for the Boston Red Sox. Often described as "the brainiest player in major league baseball," he had passed the bar exam in the off season and years later been recruited to join the OSS.

Berg had been called to Paris just before the Battle of the Bulge for a week of OSS briefings on a top-secret mission that would take him to Zurich. Sam Goudsmit was one of the briefers. It is not known exactly what was said during Berg's briefings, but Berg, an inveterate note taker, left a fragmentary scrap of paper reading "gun in my pocket," and on the next line, "nothing spelled out but—Heisenberg must be rendered *hors de combat*," a French phrase meaning "out of the battle."[2]

On December 18, 1944, Berg and an OSS officer arrived at the University of Zurich lecture hall on Ramstrasse in plenty of time for the 4:15 seminar. Encountering no security, they left their hats and coats at the door and took seats in the second row. Only about twenty people were in the audience. Standing at a blackboard in the front of the room, Heisenberg began his lecture. Berg had been briefed on nuclear fission, but this discussion on something called S-matrix theory seemed to have little to do with that.

The gun weighing heavily in his pocket, Berg listened intently, trying to decide if Heisenberg should live or die. Well into the lecture, he began taking notes: "thinish . . . Irish look . . . Heavy eyebrows, emphasize movement of that part of bony structure over the eyes . . . Sinister eyes . . . Continuous seeming quizzical smile as he talks."[3] Berg did not hear anything sufficient to justify an assassination, so he issued Heisenberg a stay of execution.

A few days later, Berg attended a dinner party hosted by Scherrer in Heisenberg's honor. Heisenberg had asked that politics not be discussed, but with news of the German attack filling newspapers and radio broadcasts, the topic was impossible to avoid.

After dinner, the guests split up into separate rooms. A young Dutch physicist later reported that he had overheard Scherrer's wife grilling Heisenberg about Nazi crimes against the Jews. "What do you say to these atrocities?" she demanded.

Heisenberg claimed that he knew nothing about the murder of Jews in Holland and France. No one believed him. Joining in the discussion, the Dutch physicist then accused Heisenberg of supporting the Nazi regime.

"I'm not a Nazi," Heisenberg answered angrily, "but a German."[4] Later, pressed to admit that Germany would likely lose the war, Heisenberg replied, "Yes, but it would have been so good if we had won."[5]

Years of life under Nazi rule had taught Heisenberg the dangers of speaking his mind. "We all had to disguise our words and actions as best we could," Elisabeth Heisenberg wrote later. "Even among acquaintances and casual friends we were guarded, and we spoke only in innuendoes."[6]

Heisenberg knew too that neutral Switzerland was full of Allied and Nazi spies. But that night in Zurich, his emotions got the better of him. That night he spoke the truth, and he spoke it publicly where everyone could hear. He wanted Germany to win the war but feared it would not. His former friends on the Allied side heard, the Gestapo heard, and in future years, historians trying to understand his motives would hear.

When Goudsmit learned what Heisenberg had said, it fueled a growing contempt for his former friend.[7] When the SS in Berlin got wind of the remark, it prepared to charge Heisenberg with treason. Such defeatist comments were punishable by death. But once again Heisenberg had a friend in a high place. That friend promised to reprimand Heisenberg, and the incident was swept under the rug.[8]

Moe Berg also heard Heisenberg's comment, and that night after the dinner, he arranged to walk with Heisenberg back to his hotel. Still deciding whether to carry out the assassination, Berg tried to draw out Heisenberg's feelings about the Nazi regime and the war, but he refused to go along. As they approached his hotel, Berg concluded that the reputed leader of Germany's atom bomb program would not have been so defeatist if he had succeeded in building an atom bomb. Heisenberg's impassioned outburst had saved his life. The two men shook hands and parted.

CHAPTER FORTY-EIGHT

BATTLE OF THE BULGE AFTERSHOCKS

The aftershocks from the Battle of the Bulge reached all the way to the United States—to Chicago, Washington, D.C., Hanford, and Los Alamos. In Chicago, Jim Schoke, soldier and amateur magician, bounded on stage for his second-annual performance at the Met Lab Christmas party. The previous year's guests had included about a hundred civilian and military personnel, half of them commissioned officers in uniform. Not this time. Schoke was puzzled because the 1944 turnout was considerably smaller, with almost no uniformed personnel. He later learned that "one of the top officers had received a tip from Washington that the Germans had succeeded in making an atomic bomb and were going to drop it on Chicago on Christmas day."[1] Word got out, and many of the would-be revelers elected to spend Christmas outside the city.

In Washington, General Groves took the German attack as a warning not to let his guard down. Gambling that Oppenheimer would solve the plutonium detonator problem, he ordered Du Pont, the company that built and operated the Hanford reactors, to accelerate plutonium production.

Standard protocol required that the irradiated uranium slugs from the nuclear reactors be stored in cooling ponds for 120 days so that the most radioactive by-products would have time to decay to safe levels before further processing.[2] General Groves ordered that the time be cut in half, and then cut again. After the cooling ponds, the irradiated slugs were transported on railcars to a remotely operated chemical separation facility where the plutonium was extracted and radioactive gases released through two-hundred-foot-tall stacks into the atmosphere. The speeded-up schedule increased the amount of plutonium produced, but it also increased the amount of radioactive gas emitted.

The health effects on employees and communities downwind of Hanford would not be known until several years later. General Groves worried that workers might quit or leak information about the risks posed by the shortened schedule, so he adopted a "less [said], the better" attitude.[3]

At Los Alamos, Oppenheimer and his scientists were already working six-day weeks to develop a plutonium detonator. Hanford's accelerated schedule pushed everyone to the limit. The relentless pressure left Los Alamos's chain-smoking scientific director emotionally drained and physically exhausted. By the spring of 1945, the already frail Oppenheimer had lost thirty pounds and weighed a mere 115. He was coughing incessantly. And, to make matters worse, he'd caught chicken pox.[4] Concerned about Oppenheimer's health, General Groves violated his own security regulations and invited Oppenheimer's brother Frank, a former member of the Communist Party, to come to Los Alamos to offer moral support.

The stress of America's atom bomb quest had whittled away at Oppenheimer; sent Glenn Seaborg, the U.C. Berkeley physicist and lead plutonium researcher, to the hospital; and sent Sam Goudsmit back home to Michigan for some much-needed R&R. Those were the known casualties. How many others managed to muscle through and avoid public detection isn't known.

But one of them checked into the Oak Ridge hospital several times complaining of severe pain in his chest and arms and always got special treatment. He was examined personally by Dr. Charles Rae, the hospital's director, and was usually discharged within hours. On the rare occasions when he was admitted overnight for observation, his EKGs were inconclusive. Dr. Rae ultimately concluded that his patient, General Leslie Groves, was suffering from atherosclerosis, a disease that restricted blood flow, and had likely caused at least one heart attack.[5]

General Groves managed to keep his heart condition secret for the rest of the war. Fortunately, the condition did not worsen. If it had, the secretary of war would have been hard-pressed to find someone qualified to replace him. Blessed with a prodigious memory, Groves kept limited records and often failed to keep his aides fully informed about what he was up to. In his autobiography, the general acknowledged that his failure to prepare someone to take over was a mistake and could have caused serious trouble if he had died or been disabled.[6]

CHAPTER FORTY-NINE

THE LAST EXPERIMENT

CBS NEWS, JANUARY 28, 1945

The pressure is on from the west and the east. Thousands of cold and tired doughboys dressed in their white snowsuits are moving like ghosts through the forest. And 600 miles to the east the same kind of cold, tired, white-clad soldiers are moving towards the west. The world's greatest military squeeze play will end when those American doughboys shake hands with those Russian soldiers somewhere in Germany.

By late January 1945, the red army had advanced to within sixty-five miles of Germany's capital, and Berliners were panicking. The German Army could do little to stop them, and the Luftwaffe had been largely destroyed. The war was lost, but Gerlach reasoned there was one last hope. If German physicists could build a working nuclear reactor, they might be able to use it as a bargaining chip to win more favorable treatment after the war.

The results from Heisenberg's previous *B-VII* experiment in the Berlin underground bunker gave Gerlach cause for optimism. His team had significantly increased the production of neutrons by surrounding the reactor with graphite and reflecting escaped neutrons back into the reactor. Heisenberg's team had finally learned what Fermi and Szilard had discovered years earlier and Professor Bothe had failed to see—that graphite, once it was purified, would make an excellent moderator. More importantly, Heisenberg had finally swallowed his pride and abandoned his design using alternating layers of uranium metal plates and heavy water. Instead, he allowed his colleague Professor Wirtz to begin assembling reactor *B-VIII*, their largest reactor to date, using Diebner's design with hundreds of small uranium cubes immersed in a ton and a half of heavy water.[1] Wirtz still considered *B-VIII* to be a subcritical assembly, so he did not build in control rods as protection against a runaway nuclear chain reaction.[2] He and his crew were working night and day to get ready for their next experiment, but time was running out.

Russian troops were now within fifty miles of the outskirts of Berlin, and Gerlach decided they had to leave. On January 30, 1945, Gerlach, Wirtz, and Diebner, assisted by German troops, loaded all their equipment and supplies into a convoy of trucks and headed south on the autobahn toward Haigerloch, where Heisenberg had set up his new lab. But halfway there, Gerlach changed his mind and directed the convoy to turn toward Stadtilm, the site of Diebner's relocated lab.[3]

Diebner's reactor experiments had produced better results than Heisenberg's, so Gerlach decided he should get Heisenberg's share of uranium and heavy water.[4] When Heisenberg found out, he was livid. The Nobel laureate had a large but fragile ego. Gerlach's decision to hand over his supplies to Diebner, a scientist Heisenberg considered his inferior, was a slap in the face. Heisenberg needed to be recognized as the best, wrote historian Paul Lawrence Rose.[5] Giving Diebner his supplies was an attack on his honor. And for Heisenberg, preserving that honor was a supreme value.[6]

With telephone communication spotty, Heisenberg and Weizsäcker decided to travel 250 miles to Stadtilm to convince Gerlach that they were close to achieving a chain reaction and should get the supplies. Setting out before sunrise on February 5, 1945, they biked twenty kilometers to a railway station in Horb where they were able to catch a train headed to Stadtilm, but the train could only take them part of the way because Allied bombs had destroyed the tracks. They then hitchhiked to another station only to find that those trains were also out of commission. Finally, they were able to telephone Gerlach, and he promised to send a car.

The car arrived hours later, but their drive was cut short when an Allied fighter plane spotted them and opened fire, forcing them to take cover in a ditch. It was night when they finally reached Stadtilm where they were greeted by blaring air raid sirens and skies dense with enemy planes.[7]

They located Gerlach the next day, and, after some intense discussion, convinced him to drive to Haigerloch and inspect Heisenberg's lab. A short while later, after allotting Diebner some of the heavy water, Gerlach ordered the remaining amount and the uranium cubes to be loaded back on the trucks that brought them to Stadtilm and rerouted to Heisenberg's laboratory in Haigerloch.[8]

CHAPTER FIFTY

DIEBNER'S LAB CAPTURED

Goudsmit was in Paris and Colonel Pash at Alsos headquarters in Heidelberg. They knew where Heisenberg and Kurt Diebner's labs were located and were waiting to see which would be overrun by U.S. troops first. When Goudsmit got word that General Patton's Third Army would soon arrive at the town of Thürigen where Diebner was located, he arranged for a friend to fly that critical intelligence to Col. Pash in Heidelberg. Pash immediately assembled a small convoy of scientists and soldiers to head to Thürigen. Goudsmit would fly in the next day.[1]

Diebner's lab was located in an old schoolhouse. When Pash and his convoy of armored jeeps pulled into the school's courtyard, the Germans inside were stunned. Pash informed them that the U.S. Army now controlled their town and ordered them to notify the town administration. Meanwhile, several combat engineers assigned to Alsos searched the premises for booby traps and inspected the building looking for Diebner's lab. They found a classroom with the basic equipment one would use in a high school science course, but nothing to suggest a well-funded, government-run project.

Continuing the search, Pash discovered a long hallway with a large garage-like door at the end. In front of the door, he noticed what looked like a block of paraffin, the kind one might use as a moderator in a nuclear reactor. When he slid the door open, he was surprised to find several men, women, and children huddled in what appeared to be a cave. They said they were residents of the town seeking shelter from possible shelling and denied any knowledge of a laboratory. But then Pash spotted several black briquettes on the ground.

"They're just coal," one of the men told him, but when Pash picked them up, he could tell they were too heavy to be coal. An Alsos scientist then took a look and confirmed they were actually pressed uranium powder, the kind used in a nuclear reactor.[2] The Germans were lying. Pash subsequently discovered a deep pit dug in the center of the cave for a reactor and marks on the walls and floor suggesting that the lab's equipment had been dismantled and removed.

Pash held the group in custody, and the next day, Goudsmit conducted the interrogation. The men and women were, in fact, laboratory staff. They revealed that Walther Gerlach, the head of Germany's atomic research program, had once been headquartered in the town, and that SS troops had recently evacuated Diebner and his key personnel, along with their equipment and supplies. They had left two days before Pash arrived.[3]

CHAPTER FIFTY-ONE

HEISENBERG'S FINAL EXPERIMENT

The shipment of heavy water and uranium that Heisenberg reclaimed from Diebner was sufficient for one last attempt at sustaining a nuclear chain reaction. Heisenberg's final experiment, number *B-VIII*, would take place in Haigerloch inside a cave at the bottom of an eighty-foot-high cliff safe from Allied bombs. The cave had previously served as a wine cellar for a nearby restaurant. Burrowing deep into the cliff, engineers enlarged the space to make room for tanks to hold the heavy water and a storage area for hundreds of uranium cubes. They installed electricity, a winch, and other equipment, and in the center of the floor, dug a ten-foot-wide pit for the *B-VIII* reactor vessel and lined it with ten tons of graphite blocks to reflect escaped neutrons back into the reactor.

On the morning of March 24, 1945, Werner Heisenberg, Karl Wirtz, and several other members of the Heisenberg team watched as the winch slowly maneuvered a lid into position above the reactor. Suspended from the bottom of the lid were 664 small uranium cubes weighing a total of one and a half tons.[1]

Once the lid was lowered, technicians bolted it closed and pumped ordinary water into the pit. They then inserted the neutron source, five hundred milligrams of radium-beryllium, through a pipe in the lid. When all was ready, they pumped the precious heavy water—one and a half tons—into the reactor. As the heavy water level rose, a Geiger counter registered the increased neutron production. Every few minutes, Heisenberg ordered the flow to be stopped so he could check the readings. He had conducted several full-scale experiments before, but none had gone this well. The rate of neutron production doubled the previous best mark, and then tripled and quadrupled it. The normally reserved scientists could barely contain themselves. The reactor moved closer and closer to criticality with neutron production numbers six, seven, and eight times Heisenberg's previous record. And as it did, the euphoria that had greeted the initial success began to dissipate. Heisenberg had never experienced a reactor this efficient. The neutron multiplication now reached a level ten times greater than any previous results.[2] He believed that the *B-VIII* reactor would stabilize at a high temperature due to an increased absorption of neutrons.[3] As a result, Professor Wirtz had not included control rods to prevent a runaway chain reaction. Now they wondered if they had made a terrible mistake.

THE GRIFFIN

Walther Gerlach was in his Berlin office on March 24, 1945, when he got a call that the reactor was about to go critical. Excited, he called his good friend Paul Rosbaud to come immediately. Rosbaud, an Austrian-born metallurgist and scientific advisor to the publisher of the scientific journal *Die Naturwissenschaften*, was well known and highly respected in the German scientific community.

When Rosbaud arrived at 1:00 p.m., Gerlach could hardly contain himself. "*Die maschine geht!*" he proclaimed. "The machine works!" In the future, there would be no need for petroleum. Everything would be run on atomic energy.[1] Rosbaud knew Heisenberg was testing a new reactor, and he knew that seven previous tests had failed to reach criticality, so he was shocked by the sudden turn of events. Once Gerlach calmed down, Rosbaud understood what had really happened. The caller had told Gerlach the new design using uranium cubes had worked. The neutron multiplication had increased dramatically. But it hadn't increased enough to sustain a nuclear chain reaction. The reactor had not gone critical.

Rosbaud didn't let on, but he was secretly relieved. He had great respect for Heisenberg's brilliance but couldn't stand his arrogance.[2] Rosbaud was also fervently anti-Nazi. But his relief had little to do with his personal feelings. Unbeknownst to his German friends, Paul Rosbaud, code-named "the Griffin," was a British agent and, according to Britain's daily newspaper the *Guardian*, "reputedly the longest-serving and best-placed spy working for Britain during the second world war."[3] He had surveilled Heisenberg's goings and comings and tracked German progress in building an atom bomb since early in the war.

Rosbaud's regard for Britain and its people dated back to World War I when his Austrian Army unit was taken prisoner by British troops and treated very humanely. He would later write, "My first two days as a prisoner under British guard were the origins of my long-time Anglophilia." More recently, in 1938, the head of the British passport office in the Berlin Embassy (who was station chief for Britain's foreign intelligence service) had helped Rosbaud's Jewish wife and daughter emigrate to England.[4] That experience led Rosbaud to sign on as a British spy.

Rosbaud's intelligence contacts came primarily from the world of scientific publishing. But Wernher von Braun, the architect of the German V2 rocket program, also gave an unwitting assist. At the beginning of the war, he had invited Germany's best technical and engineering minds to come together at Peenemünde to help brainstorm ideas for how the V-2 program might develop a long-range guided missile. Rosbaud was not invited, but he had friends who were. Those friends subsequently became unsuspecting intelligence sources.[5] With the assistance of Norwegian and French resistance networks, Rosbaud reported to London almost every month.[6]

When British intelligence discovered that the theoretical physicist Pascual Jordan, Heisenberg's former collaborator, was working at Peenemünde, Rosbaud offered to help out. He despised Jordan. He considered him a Nazi fanatic and a buffoon. Masking his real feelings, he arranged to have Jordan invited to a cocktail party. The party turned raucous, and an attractive woman plied Jordan with liquor and questions about Peenemünde. His answers provided valuable information about both the V1 and V2 flying bombs.[7]

Rosbaud also helped Lise Meitner, Otto Hahn's Jewish colleague who discovered nuclear fission, escape Germany following the passage of new anti-Semitic laws. And he played a key role in alerting the world to the possibility of an atom bomb. In December 1938, he arranged for the scientific journal *Die Naturwissenschaften* to rush publication of Hahn and Strassmann's research, which touched off a global rush to perform nuclear fission research.[8]

HUNTING HEISENBERG II

BBC RADIO, MARCH 18, 1945
FREDDY GRISEWOOD REPORTING

The Allies are across the Rhine. Troops under General Hodges established a beachhead on the east bank south of Bonn yesterday afternoon. A Reuters report from the First American Army says that the crossing was made at Remagen, which is about twenty-five miles south of Cologne.

By March 1945, after Allied troops crossed the Rhine, General Groves was finally convinced that Germany did not have an atom bomb. As a result, Alsos priorities shifted. Instead of gathering intelligence, Alsos focused on making sure that no German scientist, no documents, uranium, heavy water, or equipment connected to the bomb program would fall into Soviet or French hands.[1] But implementing that shift became problematic.

At Yalta, Roosevelt, Churchill, and Stalin had agreed that postwar Germany would be divided into four occupation zones administered by the Americans, Soviets, British, and French. Most of the German scientists, equipment, and uranium were outside the U.S. zone. Auer, the metal refining company that cast uranium plates and cubes for Heisenberg and Diebner, was in the Russian zone. On March 15, 1945, the U.S. Army Air Force solved the problem with a bombing raid by six hundred Flying Fortresses that destroyed the plant. Allied forces had previously seized 1,100 tons of uranium metal hidden in a salt mine and shipped it to an airport hangar in a city outside the Russian zone.[2]

While the army removed the uranium, Pash and Goudsmit focused on hunting down German scientists. On March 30, the Alsos mission occupied the physics laboratory of the Kaiser Wilhelm Institute for Medical Research in the city of Heidelberg.[3] There they captured a colleague of Heisenberg, the nuclear physicist Walther Bothe, whom Goudsmit knew personally. Goudsmit felt awkward meeting him as an enemy and wasn't sure how to proceed. Bothe set the tone, shaking Goudmit's hand and greeting him warmly. (Shaking hands was not allowed under Allied regulations, which prohibited fraternization with the enemy). Bothe had previously been interviewed by army personnel and was pleased to have someone he could talk with who understood physics. He took Goudsmit to visit his laboratory and talked about his research. He stopped short when Goudsmit raised the subject of the war.[4] But he did volunteer that the German atomic program was very small, involving roughly two dozen people.[5]

Bothe was reluctant to talk, but three other German scientists were not. They revealed that Alsos's number one target—Heisenberg—was at his lab in Hechingen in southwestern Germany, and his *B-VIII* reactor was nearby in Haigerloch. Pash was champing at the bit to go after Heisenberg, but Heisenberg and his colleagues were in the French Occupation Zone, and French troops were rapidly approaching.

Given the delicate political situation, General Groves dispatched one of his men to Europe for a high-level meeting with military commanders. In coordination with Pash and Goudsmit, they planned a quick commando operation. Parachutists of the 13th Airborne Division would land near Hechingen and secure the area. A small group of Alsos personnel including Goudsmit and Pash would follow, arriving by air and take Heisenberg and colleagues into custody. (The record suggests they would parachute in, but that's not clear). Meanwhile, the 13th Airborne would seize a nearby airfield so that a transport plane could land and evacuate everyone. Code-named Operation Effective, it was scheduled for April 22 but then cancelled because the troops and plane were needed elsewhere.[6]

Instead, Pash put together an ad hoc ground operation. Departing from their normal procedure, which had the Alsos scientists wait and follow front-line combat troops, Pash elected to operate behind enemy lines. On April 23, 1945, an Alsos convoy of two armored cars, four machine gun–equipped jeeps, and troops borrowed from an engineer combat battalion moved into territory recently abandoned by retreating German forces. They expected fierce resistance, but as they approached Haigerloch, they were greeted by white sheets and pillowcases hung from windows and a flagpole signaling surrender.[7]

Alsos agents immediately set about interrogating the residents, looking for Heisenberg's laboratory. They pointed to an eighty-foot-high cliff that towered over the town. At its base was a padlocked steel door and the concrete entrance to what used to be a wine cellar for a nearby restaurant. Townsfolk suspected it now housed some sort of secret project because they were not allowed near it. A piece of paper glued to the door gave the name of the cave's manager. Pash immediately sent someone to locate him. When he arrived and hesitated to unlock the door, Pash ordered one of his men to shoot the lock off and the manager as well, if he got in the way, and the manager complied.[8]

Inside they discovered their holy grail, the outer shell of Heisenberg's *B-VIII* reactor. "I made a strong effort to conceal my emotion," Pash would later write. His colleagues didn't try. There were smiles all around. When the Alsos scientists saw the reactor, they were shocked at the lack of radiation shielding.[9] Heisenberg was extremely fortunate it didn't go critical. If it had, he and his colleagues would have been in serious trouble. Pash did not find any uranium or heavy water. Heisenberg had ordered his team to hide them. But Pash would later discover some six hundred uranium blocks buried in a farmer's plowed field and several oil drums full of heavy water at the bottom of a cesspool.[10]

America's scientists entered the race, convinced that German science was the best in the world. The Germans believed it. The president of the United States believed it. Fear of German scientific superiority had driven President Roosevelt to authorize the construction of the atom bomb. Germany's scientific prowess had kept General Groves, Oppenheimer, Fermi, and Goudsmit awake at night. But Werner Heisenberg, Germany's most brilliant theoretical physicist, a man Allied intelligence called "the brain," had not come close to building an atom bomb. He hadn't even built a working nuclear reactor.

Pash assigned a team of engineers to dismantle and blow up what was left of the reactor so that no other country could benefit from Germany's nuclear fission research. Wasting no time, he then led a small convoy of armored jeeps and other vehicles to Hechingen ten miles away. Their destination: an old textile mill where their number one target Werner Heisenberg and his group of fifty scientists reportedly had their laboratories.

Upon arriving, the lead vehicles of the caravan took up tactical positions and set up a command post while teams of Alsos personnel combed the halls and laboratories inside the textile mill, looking for scientists on their target list. In a short while, they had taken into custody over a dozen scientists, including Wirtz and Weizsäcker. But they couldn't find their number one target. By the time they located his office, he was gone.

On Heisenberg's desk they found a photo of him. It was taken in the summer of 1939 on the campus of the University of Michigan, a prewar keepsake perhaps of happier times. He was smiling for the camera. Standing alongside, also smiling, was a man the Alsos personnel knew very well, Sam Goudsmit. Goudsmit wasn't there when the photo was discovered, but he certainly heard about it afterwards and had to endure considerable good-natured ribbing as a result.[11] Four days earlier, Heisenberg had left his staff in the basement of the textile mill and set out in the middle of the night by bicycle to join his family in Urfeld.

It had been a good week. Alsos had rounded up all but three of the scientists on its priority target list, and two of the three, Kurt Diebner and Walther Gerlach, were captured a short while later. Goudsmit had plenty of reason to celebrate. But, for him, the good news was outweighed by the bad. On April 25, he and Colonel Pash led an Alsos team to the town of Tailfingen where they captured Otto Hahn. While interrogating him, Goudsmit departed from the business at hand and inquired if Hahn had any news about his parents who had been arrested by the Gestapo. Hahn said "yes." They had been killed in a Nazi concentration camp.[12]

CHAPTER FIFTY-FOUR

CRITICAL MASS

In the spring of 1945, Oppenheimer estimated he had enough enriched U235 to fuel one atom bomb. The crash effort to invent an implosion detonator had paid off. And two and a half years after he signed on as the Manhattan Project's scientific director, an exhausted Oppenheimer was at last nearing the finish line. But there remained one more hurdle. He needed an answer to the questions General Groves had posed to Fermi and the Met Lab scientists in Chicago more than two years earlier: How much plutonium (or U235) was required to sustain a chain reaction and trigger an explosion? What was their critical mass? Oppenheimer's brain trust had come up with estimates for both bombs, but they couldn't confirm those numbers until they had enough enriched U235 and plutonium to run a test.

Responsibility for conducting those tests fell to the Critical Assemblies Group, a small team of courageous experimental physicists at Los Alamos. It was dangerous work that involved assembling increasingly larger masses of atomic fuel and running the risk of a runaway chain reaction. Physicist Richard Feynman called criticality testing "tickling the tail of a sleeping dragon," and the label stuck.[1]

The team conducting the criticality tests for U235 was led by Otto Frisch, now relocated from Birmingham University in England to Los Alamos. Frisch came up with a plan to determine criticality by slowly assembling small bars of U235 into a cube and monitoring the production of neutrons. Soon, two hundred Los Alamos chemists and metallurgists in coveralls and rubber gloves sitting behind protective shields were mixing U235 slurry with plastic and molding it into dozens of 1.5-inch bars.[2]

In preparing for the criticality test, Frisch had constructed a U-shaped wall of metal blocks on the workbench in his lab. He began by gently stacking the U235 bars inside the wall and assembling them into a cube. Each new bar caused a light by his side to flicker red, signifying a splitting of nuclei and an emission of neutrons. As the cube grew, the neutron production increased, the wall reflected escaped neutrons back into the cube, and the light flickered faster, signaling an approaching critical mass.

Concentrating on his work, Frisch failed to notice that he was leaning over the cube. Suddenly, his body began reflecting escaped neutrons back into the cube as well. The flickering sped up. The dragon began to stir. The light turned solid red, and the chain reaction approached criticality. Frisch caught the change in the nick of time and, with a sweep of his arm, scattered the bars on the floor, narrowly averting disaster.[3]

Undeterred by his near-death encounter, Frisch set out once again to deliberately provoke the dragon. In Omega Canyon, a distance from Los Alamos, he constructed a six-meter-tall structure resembling an oil well derrick, encircled it at table height with blocks of U235 compound, and hung a small steel box containing more uranium compound at the top. Working by remote control, he released the box, which slid down guide wires past the U235 blocks, and for a split second, the two formed a critical mass. The sleeping dragon awoke and spit out a burst of neutrons and a blast of heat.[4]

After measuring the reactivity, Frisch repeated the experiment over the next several days using larger and larger slugs. The reactions lasted less than a second, but the data they generated was invaluable. For the first time, Oppenheimer and Groves had empirical evidence of an explosive chain reaction.[5]

The criticality data enabled Oppenheimer to mold a critical mass and build the uranium bomb. But one bomb was not enough. A backup was required in case the first one malfunctioned. Oak Ridge would not have enough enriched U235 for a second bomb until the end of the year, so the reserve bomb had to be fueled by plutonium. But one plutonium bomb wasn't sufficient either. Before the plutonium bomb could be used, its detonator, requiring the simultaneous detonation of thirty-two conventional explosives within a millionth of a second, had to be tested too. That kind of precision couldn't be guaranteed without a full-scale test.[6] Oppenheimer now needed three bombs.

CHAPTER FIFTY-FIVE

A CALL FROM THE SECRETARY OF WAR

General Groves had done everything he could to keep the bomb secret—secret from the Germans, Japanese, and Soviets, of course, but also from snooping reporters and meddlesome politicians.[1] Henry Stimson, the courtly seventy-eight-year-old secretary of war, was a staunch ally in that effort, burying the Manhattan Project's multimillion-dollar appropriations within the War Department budget under innocuous headings like "Expediting Production."

Stimson was always deferential in his dealings with Groves. When his secretary called to arrange a meeting, she would ask if it was convenient for the general. Her call early in April 1945 dispensed with such niceties. "The secretary would like to see you at once, and he doesn't care what you are doing. It's extremely important," she said.[2]

Alarmed, Groves hurried to the secretary's office. There, Stimson told him that a Michigan congressman, Albert Engel, had learned about a very large top-secret project being constructed by the army outside Knoxville, Tennessee, and demanded to know what was going on. Stimson and the army chief of staff had briefed the leadership of the House of Representatives and the Senate on the Manhattan Project. And they had all agreed to keep it secret.

But Engel was not a member of the leadership, and he refused to go along. Convinced that the construction was some sort of government boondoggle, he threatened to launch an investigation. To placate the congressman, Stimson agreed to take him and other suspicious lawmakers to Tennessee, but first he asked Groves to give him a tour.[3]

Groves met Stimson in Knoxville and drove with him to the sprawling fifty-nine-thousand-acre uranium processing site officially named the Clinton Engineer Works. After a quick stop at the government guest house to drop off the secretary's bag, they drove by Oak Ridge—a community of single-family homes, dormitories, schools, cafeterias, churches, recreational facilities, a hospital and, according to a postwar tally, thirteen supermarkets, nine drug stores, and seven movie houses. Chiseled out of the Black Ridge foothills and constructed at a breakneck speed in less than three years, it was home to some seventy-five thousand workers who ran the separation processing plants.[4]

Then they headed out of town along a two-lane gravel road that took them past a billboard reminding passersby: "The Enemy is Looking for Information. Guard your Talk." As they traveled through hills and valleys, they could see in the distance an occasional ramshackle house or barn abandoned by the families evicted to make room for the bomb facilities.

Just over ten miles from Oak Ridge, they came upon a second town with acres and acres of trailers lined up on one side of a small commercial center and rows of barracks along the other. Named Happy Valley, it was built as temporary housing for fifteen thousand construction workers.

The plant Groves wanted to show Stimson lay just beyond the valley on the other side of a forested ridge. Before the two men could see the plant, they could hear it: an eerie high-pitched hum that sounded like a large swarm of bees. The sound grew louder the higher they climbed. When their car finally emerged from the trees at the top of the ridge, Secretary Stimson gasped. There, in the middle of nowhere, was a mammoth four-story building standing on land that just two years earlier had been wilderness. Shaped like a giant U, it stretched more than half a mile and was a quarter mile in width. The humming sound came from the uranium gas flowing through the hundred miles of pipe inside the building. Before them lay what journalist and bomb historian Stephane Groueff called "the most fantastic technological complex ever built in this country."[5]

Once inside, Stimson found himself in a fluorescent-lit, windowless corridor that seemed to stretch forever. On either side were the vital organs of gaseous diffusion—miles of metal pipe, oversized pumps, and thousands of vacuum chambers.

Thanks to some inspired engineering and a Herculean construction effort, all the design problems had been solved. Installed inside each vacuum chamber—or "diffuser," as it was called—was the crown jewel of gaseous diffusion technology: a thin sheet of metal with millions of microscopic pores per square inch rolled into a tube.[6]

As the uranium gas flowed through the porous tube, the surrounding vacuum chamber suctioned the lighter U235 atoms into the diffuser at an infinitesimally faster rate than the U238. The process was frustratingly inefficient, but after several thousand passes through the diffusers and weeks of recycling, the concentration of U235 atoms gradually began to increase.[7]

While walking the first-floor corridors, Groves, Stimson, and staff barely saw another person. The only workers in sight were the plant's inspectors, dressed in white cotton coveralls and gloves, who occasionally rode by on bicycles. The gloves assured that their bare hands never touched the machinery. If they did, the oil on their skin contacting the uranium gas would be enough to trigger a violent reaction.[8]

The second floor, a mass of gauges, pipes, and wiring, was similarly unpopulated. Like so much of the Manhattan Project, the gaseous diffusion plant broke all precedent. With its one hundred miles of pipe, it was the largest automated plant of its time.

Though automated, the plant's machinery required constant attention. On the third floor, large numbers of young college and high school graduates watched over thousands of dials, pressure gauges, and thermometers, making minor adjustments while a small cadre of PhDs watched over them. The plant never stopped. Twelve thousand employees in three shifts monitored the machinery twenty-four hours a day, seven days a week.[9]

In the spring of 1945, when Stimson visited, the plant was not yet fully operational. It had been running for a little over a month, and the concentration of U235 in the gas had reached just 1.1 percent.[10] With so much riding on the success of the plant, its engineers worried constantly. It would take only a pinhole-size leak of air to cripple the operation, turning the uranium gas to powder and clogging the microscopic holes in the tubular filter. Groves's engineers had developed a technology to detect a leak but hoped they would never have to use it.

A short drive from the gaseous diffusion plant stood a second plant, the barn-like, eight-hundred-foot long, steam-driven thermal diffusion plant modeled after the one Abelson had built for the navy. It too was separating U235 atoms from U238. And it too was working agonizingly slowly, producing only minimal separation. Under the new relay race strategy endorsed by Oppenheimer and Groves, the output of those two plants was now being fed into the electromagnetic plant to jumpstart its separation process.

If the Germans were building a bomb, could the electromagnets raise the U235 concentration fast enough to beat Heisenberg? No one knew. Everything was a gamble. But a bunch of congressional investigators spreading out across Oak Ridge asking questions and summoning Manhattan Project officials to Washington for closed-door briefings wouldn't help.

After a walk through the last plant, Secretary Stimson left General Groves and headed back to Washington to try to placate the congressmen. Meanwhile, miles away in a remote corner of the Clinton reservation, several ranch hands tended a small herd of cattle. The ranch had no barn, just a small farmhouse and a large silo under an overhanging bluff. To the casual observer it might have looked like a typical hill country Tennessee farm. It was not. The silo served as cover for a machine-gun emplacement. The ranch hands were armed security guards. And in the bluff above the silo, a concrete bunker hid a walk-in vault where General Groves stored bomb-grade U235.[11]

Shortly after Stimson's departure, a Chevrolet with Tennessee plates pulled up below the bunker. The driver and passengers were all dressed in civilian clothes, but their close-cropped hair, muscled physiques, and the bulge of the shoulder holsters beneath their jackets suggested military. They'd been coming to the ranch at the same time, 10:30 in the morning, once a week since the beginning of March 1945. There they picked up a traveling case containing vials of U235 and drove to Knoxville where they boarded the 12:50 p.m. Southland train for Chicago.

On board, they locked themselves in their reserved compartment until the next morning, when they handed their precious cargo to a second group of couriers, who in turn boarded the Santa Fe *Chief* for the long ride west from Chicago to Lamy, New Mexico. There, agents met the train and drove the suitcase up the hill to Los Alamos.[12]

CHAPTER FIFTY-SIX

HEISENBERG ON THE RUN

Early in the morning of Friday, April 20, 1945, just days before Pash arrived in Haigerloch, Heisenberg had set off on what was likely the most harrowing of his wartime travels. He commandeered a bicycle, borrowed a pack of American cigarettes from friends, and headed for Urfeld 150 miles away to join his wife and children. He biked after dark to escape strafing fighter planes and spent a good part of the daylight hours hiding in ditches. On the roads, he passed gangs of starving youths and marauding soldiers, but it was the military checkpoints set up to catch soldiers who had deserted their regiments that worried him the most. Hitler had ordered his troops not to surrender, and SS units were known to shoot or hang suspected deserters on the spot.[1]

Heisenberg had written his own travel documents. When a young soldier waved him over and asked for his papers, he was terrified they might not pass muster. As he handed over the documents, he asked the soldier if he smoked. When he answered "yes," Heisenberg reached into his pocket and pulled out the pack of highly prized American cigarettes. The soldier accepted, and the bribe worked. Heisenberg continued on his way without any more cigarettes to bribe his way out of trouble.[2]

With the German Army in disarray and offering only sporadic resistance, Colonel Pash and a task force of ten men and four vehicles took off after Heisenberg but were prevented from reaching Urfeld because a bridge on the road had been destroyed.[3] Worried that Heisenberg might be killed by the marauding SS troops or captured by the French, Pash left his vehicles behind and together with three of his men hiked up the mountain. At 4:40 in the afternoon of May 2, they reached the wooded ridge. Below them lay a breathtaking panorama of a lake surrounded by high-forested mountains and a road leading down to a village.

On the left of the road stood a large hotel and an inn and farther down, homes and boathouses. Pash and his men started down the road, but before they got to the hotel, they came under fire from a small German force and killed two of them. After the attackers fled, two high-ranking German officers arrived on motor scooters offering to surrender their whole battalion, some seven hundred men. Outnumbered seventy to one, the quick-thinking Pash ordered the officers to disarm their men and bring them into town the next day, giving him time to call in reinforcements and escape with Heisenberg.[4]

After another steep climb, Pash finally found "target number one." He was sitting on the veranda looking out at the lake and waiting for them. Heisenberg invited them into the cabin and introduced them to his stunned wife and children. When Pash told Heisenberg he was taking him to Alsos headquarters in Heidelberg, he didn't resist. "He was more than willing," Pash would later write, "the more so because he had been told that Sam Goudsmit was there."[5] Heisenberg seemed pleased, but he worried about his family's welfare if he appeared to be collaborating with the enemy. Waffen SS troops had recently hung seventeen German soldiers in the woods near Heisenberg's home, accusing them of desertion.[6]

Heisenberg asked his captors to take him into custody in the town and make a show of treating him roughly. As agreed, Pash's men forced him into an armored vehicle where he was guarded by two armed soldiers and drove him to meet Goudsmit.

Twenty-seven years after losing the First World War, Germany had lost another. Millions of Germans had been killed. Much of the country had been turned to rubble. Yet, at the moment of capture, Heisenberg felt only relief, "like an utterly exhausted swimmer setting foot on firm land."[7]

The Allies had bombed Heisenberg's laboratory and home in Leipzig intending to kill him, hatched an outlandish plot to kidnap him, and sent an OSS agent armed with a pistol to hear him lecture and decide whether to assassinate him. The exhausted swimmer had no idea how close he had come to drowning.[8]

CHAPTER FIFTY-SEVEN

THE HUNT IS OVER

BBC RADIO, MAY 1, 1945, STUART HIBBERD REPORTING

London calling. Here is a news flash. The German radio has just announced that Hitler is dead. I repeat that. The German radio has just announced that Hitler is dead.

A week after Hitler's suicide, Sam Goudsmit was waiting in front of Alsos headquarters in Heidelberg when the convoy carrying Heisenberg arrived.[1] In the preceding months, Goudsmit had met with numerous German scientists helping to decide whether they would be interned as prisoners of war or freed. As Heisenberg was ushered into his office, he knew that the fate of his nemesis had already been decided. Heisenberg was simply too valuable an asset to risk his collaborating with the French or get kidnapped and forced to work for the Soviets. Goudsmit would have to take him into custody.

It had been almost six years since Goudsmit and his wife had hosted Heisenberg in their Ann Arbor home. There was so much to say that the two ended up saying very little. Goudsmit asked Heisenberg, as he had in 1939, if he wanted to come to the United States. And once again the answer was no. He would be needed in Germany after the war, he said. Then Heisenberg proudly asked Goudsmit if he wanted to see his prototype nuclear reactor.

"He was still too impressed by his own importance," Goudsmit later wrote. "I merely thanked him for his offer and left him secure in the belief that his work was ahead of ours."[2] In his account of the meeting, Heisenberg completely misinterpreted Goudsmit's comments. "The conversations with Goudsmit . . . were as amicable as though the last six years had never taken place, and I myself haven't felt this well for years, both emotionally and physically," he wrote his wife.[3] The letter was never mailed. Elisabeth discovered it among her husband's papers after his death.

The day after the meeting, Germany surrendered, and the full horror of Nazi atrocities soon became public. Newsreels from the period showed ditches filled with rotting bodies piled one on top of another and prisoners not yet dead, waiting to die.

NAZI MURDER MILLS, FIRST NEWSREEL PICTURES OF NAZI DEATH CAMPS UNIVERSAL NEWSREEL

For the first time America can believe what they thought was impossible propaganda. Here is documentary evidence of shear mass murder.

Among the dead were Enrico Fermi's father-in-law, an admiral in the Italian navy, and Goudsmit's parents. With typical German efficiency, the Nazis had recorded the date of their murder. It was February 11, 1943, Isaac Goudsmit's seventieth birthday. Heisenberg knew Goudsmit's parents. He had met them in 1925 at Goudsmit's home in the Netherlands. When the Gestapo arrested them, a friend alerted Heisenberg, hoping he would intervene. Heisenberg responded by sending a letter to Nazi authorities in Holland saying that Goudsmit had always been a friend of the German people and "He would be very sorry if . . . [Goudsmit's] parents would experience any difficulties in Holland."[4] Goudsmit considered the letter woefully inadequate and could not forgive Heisenberg. Nor could he forgive himself.

After the German surrender, Goudsmit returned to the Hague and to the house where he grew up. The childhood home he remembered so fondly was now a shambles. Its wooden window frames, doors, and staircase had all been removed to be used as firewood to heat Dutch homes during the previous winter. Broken glass from stained-glass windows littered the ground. Standing in front of the wreckage, Goudsmit was overcome with remorse over his parents' murder.

"Maybe I could have saved them," he wrote in his memoir. "After all, my parents already had their American visas. Everything had been prepared. . . . If I had hurried a little more, if I had not put off one visit to the immigration office for one week, if I had written those necessary letters a little faster, surely I could have rescued them from the Nazis in time. Now I wept with a heavy feeling of guilt in me."[5]

CHAPTER FIFTY-EIGHT

USAAF ATTACK ON TOKYO

"THE FIRST STRIKES ON IWO JIMA," COMBINED AMERICAN NETWORKS, FEBRUARY 19, 1945, ARTHUR PRIM REPORTING

A powerful task force of the Pacific fleet is now attacking enemy aircraft, air bases and other military targets in and around Tokyo. Surface units of the Pacific fleet are bombarding Iwo Jima. Aircraft of the strategic Air Force Pacific Ocean areas are bombing Iwo Jima.

In February 1945, Allied naval and air forces were making steady progress closing in on Japan's five main islands, but XXI Bomber Command, based on the island of Guam 1,500 miles from Japan, was struggling. Its new B-29 Superfortress bombers, the highest-flying, longest-range bombers ever built, had orders to destroy Japan's aircraft industry, but their raids on the Nakajima Aircraft Company outside Tokyo had run into an unexpected problem.

On November 24, 1944, 110 B-29s had taken off from Guam headed toward Tokyo. After six hours of flying, they came within view of Honshu, Japan's main island, and began the climb to their attack altitude out of range of enemy ground fire. At thirty-seven thousand feet, they headed toward the iconic Mount Fuji volcano, then turned west over Tokyo toward their target. And that was when the trouble began. The B-29s were equipped with the sophisticated top-secret Norden bombsight, a kind of analog computer for aiming bombs. The sight enabled bombardiers to make fine adjustments for aircraft speed, wind speed, and direction, assuring that their bombs were aimed accurately.[1] But the B-29s ended up damaging only 1 percent of the plant.

The frequently cloudy skies over Japan accounted for part of the dismal record. But there was another, far more significant problem. The bombardiers couldn't get the telescope on their Norden bombsights to line up with the target. Something was happening high in the skies over Japan that the B-29 crews had not experienced anywhere before. They ultimately discovered the ground speeds they were typing into their bombsights were incorrect. One bombardier later reported he had entered a speed of 340 miles per hour only to discover that his actual speed was 480 miles per hour. It appeared that a powerful hurricane-force tailwind was driving the bombers faster than anybody realized. The miscalculation resulted in the bombs being dropped several miles past the target. When a bombardier on one of the B-29s tried to explain to his superiors what had happened, it didn't go over well. "When they debriefed us, they gave us the third degree. They wouldn't believe us. 'There's no such thing as a 140-mile-an-hour wind up there over Japan. You're lying. You didn't make it over the target.'"[2]

General Haywood Hansell, the head of XXI Bomber Command and a strong proponent of high-altitude precision bombing, ordered four more raids on the plant, all with the same results. No matter how many adjustments they made to their bombsights, they couldn't hit the target. The bombs impacted in the general area but did almost no damage to the plant. After the war, meteorologists would discover bands of extremely strong wind in the northern and southern hemisphere that coursed from west to east in the upper atmosphere near the poles in the summer and moved toward the equator in the winter. They called them the "Jet Stream." In the winter of 1944 and early spring 1945, the northern Jet Stream was positioned directly over Japan.[3]

XXI Bomber Command's poor performance prompted General Hap Arnold, head of the Army Air Forces, to replace General Hansell. His departure signaled an end to the so-called precision bombing doctrine which Hansell had helped draft. General Curtis LeMay, Hansell's replacement, decided it was time for serious changes.

Under pressure to destroy the enemy's will to fight, General LeMay abandoned precision bombing almost entirely. He scheduled nighttime raids and ordered his pilots to fly *below* the clouds at altitudes between five and nine thousand feet—altitudes unheard of for the giant B-29 Superfortresses—leaving them exposed to ground fire. He stripped out all defensive armament except a tail gun to reduce weight so that the planes could carry a double load of napalm-fueled M-69 incendiary bombs. And he expanded the target-selection criteria to include civilian as well as military targets. The general knew his plan would be controversial, so he cabled it to his boss General Hap Arnold at a time when he knew that Arnold would be out of the office.[4] For his first raid, he chose a twelve-square-mile region of Tokyo with many small factories wedged in among thousands of working-class homes.

During the war in Europe, the USAAF had considered targeting civilian areas extreme and inhumane. Less than 4 percent of U.S. bombs in Europe were aimed at civilians.[5] But as the war progressed, safeguards adopted to limit civilian casualties eroded. American bombers were permitted to operate blind in cloudy weather, releasing their bombs without eyes on the target. They were also authorized to add more incendiaries to their normal mix of bombs, both of which led to increased civilian casualties.[6]

The U.S. Army Air Force had played a support role in two of the RAF's most controversial raids targeting civilians. During the 1943 Hamburg raids, the USAAF conducted daytime raids on strategic targets, but they also assisted the RAF by setting fires as beacons for the RAF's nighttime raids targeting commercial and residential areas.[7] And on February 14, 1945, after two waves of RAF bombers had set the historic city of Dresden, Germany, on fire with hundreds of incendiary and high explosive bombs, USAAF B-17s launched a second attack. Their intended targets, oil refineries and a railroad marshalling yard, were obscured by smoke and clouds, so they blindly released their bombs on the city, adding to a growing firestorm. The widely condemned bombing by the two air forces set large expanses of the city ablaze, killing an estimated twenty-five thousand people and decimating a city many considered an architectural and cultural jewel.[8]

Three weeks after the Dresden attack, General LeMay's XXI U.S. Bomber Command took the killing to a level never imagined—the highest number killed and largest destruction of property of any bombing raid of World War II.[9] On the night and early morning of March 9–10, 325 B-29 Superfortress heavy bombers took off from bases on the islands of Tinian and Guam in the Pacific and flew 1,500 miles to their target, one of Tokyo's most densely populated areas, and released a cascade of incendiary bombs.

On the ground, thousands of small fires merged into large fires which merged into infernos and then powerful firestorms, incinerating large swaths of the city's wooden structures and exacting a horrific human toll: an estimated 83,793 dead and 40,928 wounded.[10]

The March 9/10 raid marked a milestone on the road to Hiroshima. USAAF precision bombing doctrine had been abandoned.[11] The new doctrine allowed targeting civilians, and, as Allied deaths mounted, it raised hopes that an invasion of the Japanese home islands might not be necessary. Currently in the planning, Operation Downfall would deploy an invasion force fifteen times larger than the one that came ashore at Normandy.[12] General Douglas MacArthur predicted, "an assault on Japan would be the greatest bloodletting in history."[13] Estimates for the number of American casualties ranged as high as one million men.[14]

CHAPTER FIFTY-NINE

CRITICAL MASS II

The convoy—three large panel trucks sandwiched between a lead and rear radio-equipped military police car—arrived early in the morning.[1] Each vehicle had two guards armed with shotguns, revolvers, and a submachine gun.[2] Their destination: a vault dug deep in the side of a mountain in Hanford, Washington. As the armed guards watched, soldiers emerged from the vault carrying dozens of wooden boxes and set them gently on racks inside the trucks. Each box contained a stainless steel flask with eighty grams of a bluish-green slurry: plutonium. No one except for an officer in the lead truck knew what they were.

From Hanford the five-vehicle convoy traveled seven hundred miles to Fort Douglas in Salt Lake City, a drive lasting more than ten hours. There they met a second convoy of soldiers who transferred the precious cargo to their trucks. The first returned to Hanford; the second headed to Los Alamos. Neither knew the full particulars of the journey—where the cargo came from and where it ended up.

The trucks delivered the plutonium slurry to the D Building at Los Alamos. There, the metallurgists transformed it into metal so that the Critical Assemblies Group could run tests to determine how much plutonium was required to sustain a chain reaction. Louis Slotin, a brilliant thirty-three-year-old Canadian experimental physicist and engineer, directed the research. Fermi had hired him at the Met Lab in Chicago and later brought him to Los Alamos. An avid scientific investigator, Slotin welcomed the challenge and shrugged off the risk of a runaway chain reaction, earning a reputation among his fellow scientists as a cowboy.[3]

He began the testing with small plutonium spheres ranging in size from 0.90 inches to 2.75 inches and worked his way up to what was estimated to be a full bomb-size critical mass divided into two separate hemispheres. In prepping for the final test, he designed a hoist with a cradle at the bottom to hold one hemisphere flat side up and an adjustable arm at the top to hold the second hemisphere flat side down. The hoist would enable him to perfectly align the two hemispheres and then, in tiny millimeter increments, lower the top half until the two were almost touching and the dragon stirred.[4] The experiment required a steady hand and lightning reflexes to cut the expected chain reaction short. Based on that brief burst of energy, Slotin's colleagues would determine the exact size of plutonium's critical mass. That measurement was vital because unlike the U235 bomb, which brought two subcritical masses together to trigger an explosion, the plutonium bomb contained just one subcritical mass. There was no margin of error. The slightest miscalculation could cause the plutonium bomb to reach criticality while it was being assembled.[5]

Slotin was a favorite of Oppenheimer.[6] On the days that he conducted criticality tests, Oppie would often show up to lend moral support. Fermi, on the other hand, often chose those days to go hiking.

On June 12, 1945, as Slotin prepared to run his final lab test on plutonium, the U235 bomb was ready to go, but the plutonium bomb still needed a full-scale test using the implosion detonator. And there was no time to waste.

President Roosevelt had died of a heart attack in April, and his vice president, now president, Harry Truman was scheduled to meet with Prime Minister Churchill and Soviet Premier Joseph Stalin in Potsdam, Germany on July 17. Their meeting would decide the future of occupied Germany and other issues arising from the war. Truman was suspicious of Stalin's intentions, and a successful test would strengthen his hand in the negotiations. The test had to be conducted before the July 17 meeting.

Back in the lab, Slotin loaded both plutonium hemispheres into the hoist. He had practiced his next moves many times using stainless steel hemispheres instead of plutonium, but this time he was doing it for real. A bright light illuminated the lab where he and Harry Daghlian, his twenty-four-year-old assistant, and a few curious observers gathered. Slotin stood in front of the hoist's hand-operated sprocket wheel, which lowered the top hemisphere into place. Daghlian stood a few feet behind him, ready to call out the radiation readings and the shrinking distance between the two hemispheres.

Everyone held their breath. The only sound came from the clicks of a Geiger counter. Slotin reached for the wheel and gave it a slight turn, lowering the top hemisphere a few millimeters. While Slotin trained his eyes on the hoist, Daghlian called out the distance between the two hemispheres and the rads, the amount of radiation emitted. Slotin gave the wheel another turn, and then another. The clicking of the Geiger counter increased. The space between the two hemispheres closed.

"Four millimeters, approaching critical," Daghlian announced. A moment later, the dragon stirred, and Daghlian yelled, "Critical!" With that, Slotin spun the wheel backwards, raising the hemisphere and silencing the dragon.[7]

The testing would continue even after President Truman's July 17 meeting. And with each assembly, the odds of something going wrong increased. Indeed, a year later, both Slotin and Daghlian had fallen victim to lethal doses of radiation from the aroused dragon.

While Los Alamos pushed to assure a successful full-scale test, a group of scientists at Chicago's Met Lab headed by Nobel laureate James Franck worked to ensure the bomb would never be used. Germany's surrender had nullified the threat of a Nazi atom bomb, and using it against the Japanese would almost certainly trigger an international nuclear arms race. Under the circumstances, Franck and colleagues recommended to an interim committee authorized by President Truman and chaired by Secretary of War Stimson that the United States arrange a demonstration—explode the bomb in an uninhabited area and invite representatives from nations around the world to bear witness.

Oppenheimer and Fermi opposed the idea. After Germany's surrender, they had redoubled their efforts to build the bomb. "We can propose no technical demonstration likely to bring an end to the war," they wrote.[8] Looking beyond the war's end, Oppenheimer argued that the best way to halt an international race to build atom bombs was to actually use it in the war. A staged demonstration couldn't possibly capture the horror the bomb would unleash on the population of a vibrant, bustling city. The whole world needed to understand that this was not merely a more lethal type of conventional bomb, but a weapon of mass destruction. One bomb could level an entire city and kill tens of thousands of people. The more horrifying the demonstration, the greater the likelihood that nations would agree to international controls. And for that reason, the United States had to use the atomic bomb in an actual war—a war against the Japanese.

General Groves opposed the demonstration as well, but as Robert Jay Lifton and Gregg Mitchell, authors of *Hiroshima in America*, argued, Groves was worried less about international repercussions than the impact on his career. He pictured himself in a packed Washington, D.C. hearing room struggling to explain to headline-seeking senators how he'd wasted two billion taxpayer dollars on a weapon that was never used.[9]

Ignoring Oppenheimer and Groves's objections, the Franck committee attempted to present their petition to President Truman, but before it could reach the president, the army got hold of it, classified it, and it was not seen again until after the war.[10]

CHAPTER SIXTY

PREPARING FOR INVASION

> **"BITTER BATTLE FOR OKINAWA NEARING END," *DES MOINES TRIBUNE*, JUNE 20, 1945**
>
> The 82-Day Battle for Okinawa was ending in complete American victory today. Many enemy soldiers, in a familiar pattern of last-ditch despair, committed suicide by hurling themselves from the 100-foot cliffs onto the jagged rocks lining the shore.

The closer American forces got to Japan, the higher the death toll.[1] Okinawa would turn out to be the bloodiest battle of the war in the Pacific: 49,000 American casualties, with 12,000 deaths.[2] The victory brought Allied forces within four hundred miles of the Japanese home islands. From there, they would launch their final invasion planned for Kyushu, Japan's southernmost island. Allied commanders knew it would be an extremely costly battle. The Japanese military had correctly anticipated U.S. plans and had increased their forces' strength there to 240,000 men from 40,000.[3] Intercepts of Japanese diplomatic traffic revealed that they were preparing an air strategy "based on total suicide air attacks."[4]

Prior to the Allied victory on Okinawa, the War Department had published gruesome photos and graphic accounts of the infamous 1942 Bataan Death March during which Japanese soldiers beat, starved, and decapitated thousands of American and Filipino prisoners of war. Those horrific reports and images further fueled a deep hatred for an enemy whom many Americans viewed as brutal, maniacal, and a subhuman race.[5] After three and a half years of fighting and the Japanese refusal to surrender, the Allies were primed to take the war to the Japanese people. And they had just the weapons to do it: the long-range B-29 Superfortress bomber, an M-69 napalm bomb specifically engineered to destroy Japan's wooden buildings, and, if Oppenheimer and Groves could deliver, an atomic bomb.

CHAPTER SIXTY-ONE

TRINITY

With the date for the Trinity test set, work crews headed out into the New Mexico desert to begin preparing the site. Two hundred and thirty miles from Los Alamos and eighteen by twenty-four miles in size, it had previously been used by the U.S. Army Air Force as a practice bombing range.[1] Unfortunately, no one alerted the bomber crews the Corps of Engineers soldiers would be there. When one crew spotted the lights of the Trinity camp during a nighttime training run, they assumed it was their target and let loose. Fortunately, the dummy bombs containing one-hundred-pound bags of sand and black powder didn't land on any of the workers, and an urgent call to the Army Air Force resolved the problem.[2]

Soon scores of Corps of Engineers personnel spread out across the desert, paving a road to the site, stringing electrical and telephone lines, and rigging the steel tower to hold the bomb. Meanwhile, other crews began constructing a hardened control bunker from which Manhattan Project scientists would detonate the bomb and a shelter where invited dignitaries would witness the explosion. The prevailing winds at the test site blew in a northerly direction, so the control bunker and VIP shelter were located six miles south of ground zero and the mess hall and barracks of Trinity's base camp another four miles to the south. Additional structures for cameras, instruments, and searchlights ringed ground zero six miles to the west and north.[3]

The U.S. Army Corps of Engineers work crews had not been at it for long when they discovered that the accidental bombing was a harbinger of trouble to come. The desert heat melted the insulation on the coaxial cable that connected measuring devices to remote recorders. Back at Los Alamos, the surface of the two plutonium hemispheres that made up the bomb's core developed blisters so that they wouldn't assemble properly. More snafus followed. The flight crew assigned to monitor the test from the air discovered that they were sharing a radio frequency with the Voice of America, the government's foreign broadcasting service. The two-way radios used by the project's scientists shared a frequency with a railroad freight yard six hundred miles away in San Antonio and a control tower at the Socorro, New Mexico, airport. To make matters worse, the molds for parts of the bomb's implosion detonator arrived from the manufacturer with cracks in them.[4]

The olive drab military sedan left Los Alamos at 3:00 p.m. on Thursday July 12, 1945, preceded by a carload of armed guards and followed by a car full of physicists and assembly technicians including Louis Slotin and Harry Daghlian. In its backseat was the bomb's plutonium core split into several pieces, each sealed in its own separate suitcase. The next day at one past midnight, a more conspicuous five-ton army truck departed Los Alamos under the cover of darkness with a wooden shipping crate strapped in its bed. Inside the case was the bomb's outer casing with its high-energy explosive charges. The army chose the late hour because there would be less traffic and less chance of a drunk driver crashing into the load of high-energy explosives. Traveling at a moderate speed of thirty miles per hour, the convoy took eight hours to arrive.[5]

The soldiers and scientists delivered the outer casing to a tent at the bottom of the Trinity bomb tower and the pieces of the core to a nearby ranch house previously owned by the now evicted McDonald family. The family's primary bedroom had been converted into a clean room with sealed windows and a large table covered in brown wrapping paper.

On the morning of Friday the thirteenth, Slotin, Daghlian, and six other technicians dressed in white surgical gowns spent six hours meticulously assembling the pieces of the core into two hemispheres. Unfazed by the discovery of tiny pits in their surfaces, they calmly filled them with Kleenex and went on about their business. Once finished, the assembly team delivered the core to the bomb tower where, after some difficulty, they lowered it into the bomb's outer casing and then winched the whole six-foot wide sphere draped in electrical cables to a shed at the top of the one-hundred-foot tower.[6]

The next day, July 14, 1945, two days before the scheduled detonation, Oppenheimer was at Trinity base camp when he received a coded message from Los Alamos. He had ordered his team of scientists to run a trial test of the bomb's implosion detonator using conventional explosives. The message came from the group's leader, Edward Creutz: the detonator had malfunctioned, and the Trinity bomb would likely fail as well.[7]

Oppenheimer was visibly shaken. He had overcome astronomical odds and vanquished extraordinary obstacles. Now, with the Potsdam conference just days away and the president relying on him, he feared he would fail. More importantly, he worried the bomb might not be ready in time to forestall the Allied invasion of Japan he knew was coming.[8] After huddling with his colleagues, Oppenheimer messaged Creutz telling him to double-check the results. But the bad news kept coming. A dummy test of the firing circuits that detonated the bomb blew out. The unit had been tested numerous times, and this was the first failure. The can-do optimism that had kept the Los Alamos scientists going through their most difficult trials began to give way to gloom and talk of a dud or fizzle. A Los Alamos physicist/poet captured the mood in the closing lines of a satirical quatrain: "Lo, the embattled savants stood, and fired the flop heard round the world."[9]

The bad news could not have come at a worse time. Oppenheimer was an emotional and physical wreck. Kitty had given birth to their second child, a girl whom they named Katherine. Kitty was drinking heavily and on the verge of a nervous breakdown. Depressed, she had arranged for a neighbor to care for Katherine and announced she was leaving Los Alamos and taking their four-year-old son to visit her parents. Oppenheimer's biographers Kai Bird and Martin Sherwin report that Kitty's departure left Oppenheimer "unusually vulnerable," although she did return before the Trinity test. During her absence, Oppenheimer visited Pat Sherr, the neighbor taking care of his baby daughter, but never asked to see his daughter. He also stunned Sherr by asking if she would like to adopt Katherine, explaining "I can't love her."[10]

Meanwhile, at base camp, the atmosphere, which had been tense in the runup to the Trinity test, turned vindictive. Oppenheimer lashed out at George Kistiakowsky, the head of the team developing the implosion detonator, blaming him for the problems. For him and his staff of six hundred, developing the explosive components for the bomb had been a long and grueling slog. In the spring and summer of 1944, none of the experimental results had pointed to a successful outcome, and an extremely pessimistic Kistiakowsky had predicted, "the test of the [bomb] failed . . . Kistiakowsky goes nuts and is locked up."[11] But a year later he stood his ground, betting Oppenheimer a month's salary against ten dollars that the implosion charge would work.[12]

On Sunday the fifteenth, Oppenheimer was at the test site's base camp when he received a follow-up message from Los Alamos. This time it was good news. The instruments monitoring the test had malfunctioned, not the detonator. Also, a retest of the firing circuit had performed well. And there was more good news. Sunday morning had begun with scattered showers, but Trinity's chief meteorologist assured Oppenheimer that the weather on test day would be quite suitable.[13]

When General Groves arrived Sunday afternoon, it was still raining. Oppenheimer shared the good news about the implosion detonator test. Groves was relieved, but worried about the weather. A rainstorm or high winds could carry the radiation from the explosion beyond the desert to populated areas.[14] If the showers lingered until the next morning at 4:00 a.m. when the test was scheduled, they would have to postpone.

At dusk, the weather began to change. The wind picked up. A storm was on its way. Killing time, Oppenheimer and Groves wandered over to the mess hall where they found several bedraggled scientists and soldiers nursing cups of steaming coffee and listening to Enrico Fermi. Fermi, who was there to measure radiation levels following the test, had organized a betting pool on the size of the explosion. The highest prediction came in at forty-five thousand tons of TNT. Oppenheimer predicted conservatively three thousand tons. Meanwhile, Fermi upped the ante by taking side bets. These focused on whether the bomb would ignite the atmosphere and, if so, how much of the planet would be destroyed. Groves was not amused but kept quiet, figuring a little gallows humor might help ease the tension.[15]

With Groves's help, Fermi and a colleague had managed to get hold of two army surplus Sherman tanks which they lined with two-inch-thick lead walls and equipped with an independent oxygen supply. They planned to use the tanks to take them near the test site after the explosion. Once there, they would fire off a rocket with a small claw on the end to gather a sample of dirt. A cable tethering the rocket to the tank would retrieve the sample.[16]

As the evening wore on and the weather showed no signs of improving, Oppenheimer chain smoked and paced the mess hall, getting more and more agitated. Fermi didn't help matters. Afraid that a sudden change in wind direction following the test might deluge the camp with radioactive rain, he pleaded for a postponement. Groves intervened, suggesting that Oppenheimer join him and go for a walk.[17]

At 11:00 p.m., Kistiakowsky and his team began arming the bomb while meteorologists set up a portable weather station and released helium balloons to measure wind speed and direction. Meanwhile, Groves and Oppenheimer drove to the control bunker six miles from ground zero where technicians were double-checking their equipment.

Groves and Oppenheimer didn't hear the explosion, which was more like a loud bang, while they were driving, but the base camp commandant did. Within minutes, he and a contingent of military police were racing across the darkened desert in their jeeps trying to locate the source. In the distance they could make out what looked like a tank. Eager to make sure the claw-equipped rockets would work, Fermi woke Weil, his young colleague, and convinced him to drive into the desert for a trial run. Fermi tried to explain what he was doing. The commandant wasn't interested. He gave Fermi a thorough dressing down, and they returned to camp.[18]

At 2:00 a.m., the skies opened up and rain lashed the base camp ten miles south of the bomb tower with thirty-mile-an-hour gusts while lightning and thunder raised concerns that the bomb's steel tower might serve as a lightning rod, triggering an accidental explosion.[19] Those concerns were not unfounded. In an earlier test, an electrical storm had triggered an accidental firing of a test device. Work crews had since wired the tower with lightening arresters to bleed off any electrical charge.[20]

Inside the camp, Oppenheimer and a very angry General Groves laced into Jack Hubbard, the Trinity test's chief meteorologist. "What the hell is wrong with the weather?" he asked. Clearly rattled, Hubbard suggested that they postpone the test until 5:30 a.m.

"You better be right on this, or I will hang you," Groves threatened.[21] By 4:00 a.m., the rain had stopped. At 5:00 a.m., loudspeakers began blaring "The Star-Spangled Banner," signaling that the countdown had begun.[22]

Groves left Oppenheimer at the control center and walked nearby to the main observation bunker where Fermi and three busloads of VIP observers were assembled.[23] Lying on their backs, with their feet facing north in the direction of the bomb tower six miles away, Groves, Fermi, and the others clutched the dark glass they had been issued for viewing the explosion and waited. Groves spent those final moments worrying what he would do if the bomb turned out to be a dud.[24]

Oppenheimer, still in the control center, was hardly breathing when the loudspeaker began the final countdown, and the voice shouted, "Now!" With that, a gold, purple, and violet flash "equal to several suns at midday" lit the desert as far as the eye could see. Next came a blast of air. VIPs at the observation bunker who had stood up after the initial flash were knocked to the ground. Then, more than a minute after the detonation came the sound of the explosion, a "deep growling roar" that reverberated off the distant mountains for a good five minutes.[25]

The force of the explosion was greater than anyone had imagined. It shook houses and broke windows 180 miles away in Silver City, New Mexico; the flash of light was seen 280 miles away in Amarillo, Texas.[26] In San Antonio, the noise of the blast sounded like "an airplane, freight train and thunder rolled into one."[27] The next day, United Press quoted a government press release stating that "an ammunition magazine containing a considerable amount of high explosives" had exploded in the New Mexico desert.[28]

The test was a success. For the first time in many months, Robert Oppenheimer relaxed and accepted a hug from Kistiakowsky, who demanded he make good on their ten-dollar bet.[29] General Groves accepted congratulatory handshakes from the invited dignitaries while Enrico Fermi walked over and picked up some scraps of paper on the ground. He had taken them from his pocket a moment before the blast and let them drop. Propelled by the force of the explosion, they had landed several feet from where he stood. Fermi had used the paper scraps in an experiment to calculate the force of the bomb. "Twenty thousand tons of TNT," he announced.[30]

William L. Laurence, a correspondent for the *New York Times* who was present that day, would later write "one felt as though he had been privileged to witness the birth of the world—to be present at the moment of creation when the lord said: 'let there be light.'"[31]

The mood in the shelter six miles north of ground zero was euphoric. Thrilled to have been part of such a historic event, everyone was making ready to return to base camp, when shortly after 6:00 a.m., a loudspeaker suddenly began emitting a clicking sound. It came from a radiation monitoring station three thousand yards from the detonation site.[32] One of several such stations positioned at thousand-yard intervals north, south, east, and west of the tower, it gave warning of approaching radiation. The first clicks came in random, barely audible bursts. But then the clicking became constant and began growing louder and louder until it suddenly stopped. The welcomed silence lasted only a short while until the radiation monitor at the next station four thousand yards from ground zero started clicking.

Robert Wilson, the senior scientist among those working in the shelter, peered out the window toward the spot where the bomb tower had once stood. In the early dawn light, he could make out what was happening. Part of the mushroom cloud above the bomb tower had split off and was drifting toward them, spreading fallout—a radioactive mix of bomb debris, fission products, and radiated soil.

Hearing the ominous noise, the MPs accompanying the scientists didn't need to be told what to do. They ran out to their carryall vehicles parked side by side outside the shelter and got ready to go, but they couldn't leave. One of the scientists, Byrland Brixner, an expert in super high–speed scientific photography, and his MP driver had taken a carryall and gone to collect film from one of his cameras and had not yet returned. General Groves had issued strict orders that everyone had to leave the north shelter together after the test. He wanted to make absolutely certain that no one was left behind.

Meanwhile, the monitoring stations continued tracking the approaching radioactive cloud. At each station, the pattern repeated itself: barely audible random clicks giving way to a stream of clicks which grew louder and then quieted.

Robert Wilson had his eyes glued to his own handheld Geiger counter. Its needle had started swinging back and forth and settled past the thirty-five-roentgen mark in the red zone when the MPs outside spotted Brixner's vehicle barreling toward them and began cheering like sports fans, "Go, Brix, go!" "Go, Brix, go!" Hearing the commotion, Wilson shouted, "Let's go!" "Get out!" Grabbing their records and whatever equipment they could carry, the scientists ran outside where the MPs were already gunning their engines. They piled into the carryalls, and together with Brixner, took off, heading toward the base camp. Behind them, the radioactive cloud continued drifting toward the northeast.[33]

Just over a thousand miles away, dock crews at the Hunter's Point Naval Shipyard in San Francisco loaded three large crates containing parts for the U235 bomb on board the USS *Indianapolis*, a heavy cruiser bound for Tinian Island in the Pacific. And, back at base camp, Enrico Fermi, George Weil, and Herbert Anderson suited up in surgical gowns and climbed in their tanks for the ride to ground zero to collect their radioactive soil samples. Anderson took the lead; Fermi and Weil followed.

As they drove off, Trinity's problem of shared radio frequencies cropped up again. A taxi service in Carrizozo, New Mexico, was using their same frequency, so they had to communicate in code. Then, about a mile outside base camp, Fermi's tank broke down. Angry and frustrated, he, Weil, and their driver hiked back to camp while Anderson continued on.

Anderson took his tank as close as he dared—about half a mile from ground zero. Through the periscope he saw what looked like a "great jade blossom." The explosion had sucked desert sand into its fireball. There, it melted, congealed, and fell back to earth in the form of emerald beads and a greenish ceramic-like glass.[34] The hundred-foot-high bomb tower had been completely vaporized. In its place was a five-hundred-foot-wide, six-feet-deep crater.[35]

The bomb blast and searing heat had destroyed many of the scientists' measuring instruments. The data they did collect suggested that detonating the bomb so close to the ground had diminished the area impacted by the blast. As a result, they decided to detonate the bomb at a high altitude to extend its destructive reach.[36] That lesson would be put to use a month later in the skies over Japan.

For the moment, however, General Groves was more concerned about the skies over northern New Mexico where the radioactive cloud was moving northeast at roughly ten miles per hour and spreading radioactive fallout across a thirty-mile-wide swath of desert.[37]

General Groves had been warned there would be fallout and had rejected the warnings calling them alarmists. But he was eventually convinced that the government could be held legally liable.[38] Thus, anticipating a northerly wind direction, Groves positioned an evacuation force of 160 enlisted men and trucks north of the test site and security officers in towns up to a hundred miles away. They would be used only in an emergency. As a further contingency, should the radiation hazard turn dire, he requested permission from the governor of New Mexico to declare martial law.[39]

Teams of scientists equipped with Geiger counters were stationed along highways and in towns to the north to measure the radiation levels. Using names inspired by the movie *Wizard of Oz* to identify themselves, they would report back via radio to medical officers at Trinity.

The chief monitors in Carrizozo, New Mexico—husband and wife PhDs from the University of Chicago, Alvin and Elizabeth Graves—went by the code name Tin Man of Station six. They had arrived at their motel the day before the test, carrying a seismograph, Geiger counter, short-wave radio, and portable generator. The elevated readings began the next day at 3:00 p.m. An hour and twenty minutes later when Tin Man radioed base camp, those readings were deep in the red. Fortunately, the radioactive cloud continued its journey north, and the radiation readings returned to safe levels.[40]

With Carrizozo out of danger, Tin Man packed up their equipment and began tracking the cloud farther north. But they soon discovered their pursuit had taken them out of short-wave range of Trinity. The cloud was still registering high radioactive readings, but they couldn't report it. Meanwhile, the radiologists at Trinity were assuring Washington that the danger had passed.

CHAPTER SIXTY-TWO

FARM HALL

In July 1945, as the USS *Indianapolis* sailed to Tinian Island where its cargo would be assembled into an atom bomb, Werner Heisenberg and nine other captive German scientists were adjusting to their new lives at Farm Hall, a Georgian country house near Cambridge, England. As guests of the British, they had all signed pledges agreeing not to try to escape and settled into a routine—breakfast at 9:00 a.m., lunch at 1:00 p.m., tea at 4:00 p.m., and dinner at 7:45 p.m.[1] In between, they took walks in the rose garden tended by German POWs, played volleyball and tennis, read books from the estate's well-stocked library, listened to the radio, played bridge, and attended scientific lectures conducted by their colleagues. The English officer in charge even arranged to bring in a piano so that Heisenberg could give recitals.[2] He could perform some of Beethoven's most difficult sonatas with technical perfection, commented one critic, but his performances were "almost completely devoid of passion."[3] All the men got along well except for Diebner, who was snubbed by Heisenberg and his colleagues.[4]

Incarceration kept the German scientists out of Soviet hands while the British decided what to do with them. But their stay at Farm Hall served another even more vital purpose. It allowed British intelligence to eavesdrop on their conversations. Unbeknownst to Heisenberg and colleagues, Farm Hall was a British safe house with microphones hidden behind the many paintings that covered the walls of the dining room, living room, and bedrooms. A bilingual team of British military monitored their conversations, preserving those of intelligence value on reusable shellacked metal disks.[5]

The disks recorded the guests debating what the Americans would do with their nuclear fission research. "I am convinced they [the British and Americans] have used these last three months mainly to imitate our experiment," one said. The scientists were anxious about the future but also trying to distance themselves from the past. One Nazi party member admitted to his colleagues that he took "the line of least resistance." Another boasted of how he had rebelled against the Nazis: "I had no picture of Hitler in my institute. They kept on coming and saying we should buy a picture of Hitler. I always said no."[6]

Heisenberg, who had refused to join the Nazi party, recounted his failed efforts to intercede on behalf of friends and colleagues arrested by the Nazis. In one case, "He had already been killed." In another, "I couldn't even find out . . . whether he was alive or dead." Historian David C. Cassidy noted that Heisenberg made about ten such humanitarian gestures, but "when Heisenberg did act, he did so as he had in past situations—offering too little, too late."[7]

Only one of the detainees accepted responsibility for his countrymen's atrocities: "We went to Poland and not only murdered the Jews in Poland, but for instance, the SS drove up to a girl's school, fetched out the top class and shot them simply because the girls were high school girls, and the intelligentsia were to be wiped out. Just imagine if they had arrived in Hechingen, drove up to the girls' school and shot all the girls! That's what we did."[8]

During their captivity, Heisenberg and colleagues got little news from home. They knew only that their country was in turmoil—controlled by occupying armies, the economy in shambles, and food and fuel in short supply. Everyone feared for the welfare of their families. Heisenberg didn't know it, but his wife and five children were better off than most. Colonel Pash had been looking out for them, making several trips to their cabin to bring groceries and other supplies.[9]

In July 1945, while Heisenberg and his fellow scientists remained captive in Farm Hall, Sam Goudsmit and a small contingent of Alsos personnel traveled to Berlin to see the laboratory where Heisenberg had done much of his research. When they got there, they found an American officer who was surprised by their interest. The Russians, he explained, had cleaned the place out. He invited them to have a look.[10]

In the yard behind the lab, they found broken lab equipment, blocks of pressed uranium oxide, and old notebooks. After wandering through the building, they went down into the basement and then the sub-basement. There they discovered the bunker fortified with six-foot-thick reinforced concrete walls and what was left of Heisenberg's Berlin laboratory. The Russians had scavenged what they could, but to Goudsmit it looked as if it had once been excellently equipped. In the center was a large circular pit that had held a heavy water reactor vessel. Metal containers and frames that suspended uranium cubes inside the vessel were scattered on the floor. The winch that held the reactor was gone, but the track that guided it was still attached to the ceiling, and remnants of a sophisticated ventilation system still hung on the walls. Airtight steel doors opened to surrounding rooms which had served as laboratories and living quarters for the skeleton crew that sheltered there during Allied bombing raids.[11]

Goudsmit, who had apparently seen one of Fermi's early reactors, was struck by how much more sophisticated Heisenberg's operation seemed. The Germans had so many advantages over the Americans at the start. Goudsmit couldn't understand why they'd made so little progress.

CHAPTER SIXTY-THREE

HIROSHIMA

Hiroshima's residents were on edge. They'd seen neighboring towns decimated by mass raids of American B-29 bombers. "B-san," they called them, Japanese for "Mr. B." But their city hadn't been touched. Rumor had it that the Americans had something special in mind for Hiroshima. Not wanting to find out, thousands of residents had taken part in planned evacuations, reducing the city population to 245,000 from a peak of 380,000.

Those who stayed behind had grown accustomed to the minute-long blasts of the air-raid sirens. They heard them at night, when the Mr. Bs flew in from the coast to a rendezvous point northeast of the city, before heading off to bomb other cities. They heard them in the morning when a single American weather observation plane flew over.[1]

But then, just before 7:30 a.m. Japan time on August 6, 1945, early warning radar operators spotted three blips on their radar screens and sounded the alarm. The B-29 Superfortresses crossed the Japanese coastline and headed to a spot fifteen miles east of Hiroshima. Assuming they were part of a reconnaissance mission targeting some other city, the operators sounded the all-clear.[2] But just after 8:12 a.m., the three blips changed direction, heading straight for Hiroshima. At 8:15 a.m., the bomb bay doors on the lead plane snapped open and released its payload, a single ten-foot-long bomb.[3] Forty-three seconds later, the energy from an atomic explosion heated the air at the burst point to sixty million degrees centigrade, producing a giant fireball. A flash of light hundreds of times brighter than the sun illuminated the morning sky. Then came the shockwave accelerating outwards at a speed of ten thousand feet per second and a force of seven tons per square meter.[4] The shockwave and fireball shattered and incinerated structures and killed almost every living being in its path. By the time the roar of the explosion caught up, several square miles of the city had been turned to rubble and flames.[5]

Within seconds, the once-bustling factory town had descended into a black hole cut off from all communication with the outside world. The Hiroshima radio station stopped broadcasting; the city's main telegraph line went dead. At the headquarters of the Japanese General Staff in Tokyo, commanders tried futilely to phone the Army Control Center in Hiroshima. Soon, confusing reports from small train stations several miles from the city started pouring into the railroad's telegraph center. There had been an enormous explosion.

The military commanders knew from the radar that only three Allied bombers had been in the area, hardly enough to take out an entire city. They knew the army had no large stores of explosives in Hiroshima. There was nothing to explain the mysterious sudden total silence. Desperate for an explanation, headquarters ordered a young army officer to fly to Hiroshima and report back. Three hours later, while still one hundred miles from the city, the junior officer spotted a smoke cloud blanketing the area where the city should have been. Only after they landed did he appreciate the full extent of the destruction. The officer radioed back: the whole city was on fire or in ruins.[6] Later estimates would put the number of people killed in the initial blast at eighty thousand. Another fifty to sixty thousand would die in the months that followed.[7]

General Groves was worried. It was 2:00 p.m. The planes had been scheduled to take off at 11:45 a.m. EST.[8] He should have heard something by now. Rather than make everyone nervous by waiting at the office, he went to play tennis, leaving orders to call him when the cable arrived. Two hours later, and no word. He returned to the office.

At 6:00 p.m., still with no word, he took Grace and Gwen, his wife and daughter, and an aide to Secretary of War Stimson to dinner at the Army Navy Club. After they were seated, a fellow officer stopped by his table. Groves shook his head and said, "Nothing yet." Forty-five minutes into the meal, a waiter approached and told him he had a call. The caller told him the planes had taken off and were on their way. That was all he knew.[9]

After dinner, Groves left his wife and daughter and returned to the office. And, finally, at 11:30 p.m. on Sunday, August 5 (1:30 p.m. on August 6 in Tokyo), the telex machine sprang to life: "Results clear-cut. Successful in all respects. Visible effects greater than New Mexico test." Announcing the results to his office staff, the normally taciturn general broke into a broad smile. Groves would not be spending the rest of his life in Fort Leavenworth for misusing government funds. The three years of hard work and billions that the U.S. Treasury had invested to invent the atom bomb had paid off. The next move was up to the Japanese: surrender or face what President Truman called "a rain of ruin from the air, the likes of which has never been seen on this earth."[10]

Well after midnight, Groves went to his private office and wrote a draft of the memo he would send to Secretary of War Stimson. He then lay down on a cot and took a nap.

The next morning, he called Grace at home to tell her to listen to the 11:00 a.m. radio newscast and that he was sending one of his public relations staff to the house. He also got word to his daughter who was taking a typing course to get home as soon as possible and messaged his son, who was at Fort Belvoir studying engineering, alerting him that reporters would likely try to contact him, and he should not say anything. (Groves's family knew nothing about the bomb until the morning radio broadcast).[11]

Oppenheimer was in his Los Alamos office shortly before 11:00 a.m. on August 6 when John Manley, one of his assistants, called to give him the news. Oppenheimer had sent Manley to Washington with orders to call him the second he got confirmation. Manley had tried to call the previous night, but General Groves would not allow it until the White House made the official announcement at 11:00 a.m. Finally, sixteen hours after the explosion of mankind's first atomic bomb, the Japanese—and the world—learned what had hit them.[12]

Shortly after Oppenheimer got the news, the Los Alamos public address system relayed it to the scientists in the technical area, among them, Enrico Fermi: "Attention please, attention please. One of our units has successfully been dropped on Japan."

That evening, Oppenheimer stood in the auditorium pumping his clasped hands over his head like a prizefighter as the crowd stomped and cheered. His only regret, he said, was that the bomb wasn't built in time to use against the Germans. Later, on his way home from a party, he encountered a distraught coworker vomiting in the bushes. The nuclear age had begun. So too had a time of reckoning.[13]

On August 6, Sam Goudsmit was combing through the ruins of Himmler's headquarters in Berlin, when an out-of-breath Alsos officer showed up and told him he had fifteen minutes to catch a plane for Frankfurt. When Goudsmit asked why, all the officer said was "orders." Goudsmit figured there would be somebody at the airport when he landed to explain, but when he arrived there was no one there, so he ended up taking a military bus to headquarters.

At headquarters, all he got by way of explanation was some lame story about officers coming from London and Washington whom he might want to talk to. Not even Colonel Pash would give him a straight answer. After dinner, during a bull session with a civilian Alsos scientist, Goudsmit learned that an A-bomb had been successfully tested in New Mexico. He had known through unofficial channels that a test was planned but was once again annoyed he hadn't been told.

Later that evening, he drove a friend to her hotel where the sleepy sergeant manning the desk was listening to a dance band on the radio. Just as Goudsmit turned to leave, the program was interrupted by a bulletin: "The United States has dropped an atom bomb on Hiroshima."[14] With that announcement, Goudsmit finally understood why he had been rushed out of Berlin. Groves feared the Soviets might try to kidnap him to get information about the bomb.[15]

FARM HALL GETS THE NEWS

"HIROSHIMA NEWS BULLETIN," BBC RADIO, AUGUST 6, 1945, FRANK PHILIPS REPORTING

Scientists, British and American, have made the atomic bomb at last. The first one was dropped on a Japanese city this morning. It was designed for a detonation equal to twenty thousand tons of high explosives.

"A DAY THAT SHOOK THE WORLD," AUGUST 6, 1945

Buildings vaporized, so too did the clothes on people's backs. Was it a bomb that should never have been dropped? Some people thought so. Others argued that it shortened the war and kept the subsequent peace."

Farm Hall, August 6, 1945. A courier arrived from London and handed an urgent message to Major Rittner, the British officer in charge. Rittner was instructed to ask the German guests to leave their rooms. Once they left, a team of technicians moved in and set about testing the microphones hidden behind paintings throughout the house to make sure they were all in working order.[1] Intelligence personnel had been translating and transcribing the secret recordings for months and periodically sending summaries to London. Now, suddenly, London wanted the transcriptions right away. Rittner was surprised by the sudden urgency, but even more baffled by the order to make certain that the Germans listened to the 9:00 p.m. *BBC* newscast.[2] After listening to an earlier radio broadcast, Rittner finally understood.

Just before dinner, Rittner called Otto Hahn into his office and broke the news about Hiroshima. "Hahn was completely shattered by the news," Rittner reported, "and said he felt personally responsible for the deaths of hundreds of thousands of people."[3] Hahn then relayed the news to Heisenberg and the others who anxiously gathered around the radio awaiting the 9:00 p.m. hour.

Long convinced of their own superiority, the Germans couldn't believe that the Americans had accomplished what they could not. "How have they actually done it?" Heisenberg asked. "I find it is a disgrace, if we cannot at least work out how they did it," he told the other scientists.[4] Asked by a colleague how much U235 was needed to make the Hiroshima bomb, he at first said "one ton" (sixteen times the actual amount), but later admitted, "Quite honestly, I never worked it out, as I never believed you could get pure [uranium] 235."[5]

The announcement caught Heisenberg completely off guard. Goudsmit had lied to him, leading him to believe that the United States was not developing an atom bomb. He was angry at Goudsmit but also humiliated. He was the lead scientist on Germany's atomic research program, yet he couldn't answer a fundamental question about the size of the critical mass.

Otto Hahn lashed out at Heisenberg: "If the Americans have a uranium bomb, then you're all second-raters. . . . You might as well pack up."

"I quite agree," Heisenberg replied.[6]

Karl Wirtz, Heisenberg's trusted colleague, blamed Walther Gerlach for the Uranium Club's failures. "He and Diebner worked against us the whole time," he argued. "In the end they even tried to take the engine away from us. . . . If we had started properly in 1939 and gone all out, everything would have been all right."[7]

While others looked for someone to blame, Gerlach fretted about how his countrymen back in Germany would react. "We will be looked upon as the ones who have sabotaged everything. We won't remain alive long there."[8]

"The impact on the ten scientists was shattering," Goudsmit wrote in his memoir about the Alsos mission. "Their whole world collapsed. At one stroke all their self-confidence was gone . . . their hope for a bright future for German science was only so much illusion."[9]

To account for their failure, the Farm Hall detainees put together a statement for the British press, saying that wartime conditions made it impossible to build a bomb. Instead, they said, they had focused on producing a "uranium engine"—a nuclear reactor—to drive machinery. While directing blame elsewhere, they seemed to forget their own failures, including Heisenberg's rejection of Speer's generous funding offer, Heisenberg and Bothe's choice of heavy water as a moderator instead of graphite, and Heisenberg's rejection of Diebner's nuclear reactor design.

One member of the Farm Hall group, Carl Friedrich von Weizsäcker, declared that he and his colleagues were morally superior to the Americans. "I believe the reason we didn't do it was because all the physicists didn't want to do it, on principles. If we had all wanted Germany to win the war, we would have succeeded."[10]

German scientists would make that claim frequently after the war. Already worrying about the opinion of future generations, von Weizsäcker continued, "History will record that the Americans and the English made a bomb, and that at the same time the Germans, under the Hitler regime, produced a workable engine."[11] With wishful thinking, he ignored the fact that the Germans had failed to build a working uranium engine and that such an engine was a first step to producing plutonium for an atomic bomb.

Heisenberg and his fellow scientists would remain at Farm Hall for five more months before being flown back to Germany. Their return was hastened by a surprise announcement from the Royal Swedish Academy of Science on November 16, 1945. Heisenberg was leafing through the just-delivered English newspapers that morning when he spotted an article stating, "The German radiologist Professor Otto Hahn, the discoverer of radio-active substances, has been awarded the 1944 Nobel Prize for Chemistry . . . His present whereabouts are uncertain, but he is believed to be at the disposal of the Western Allies."[12] British and U.S. authorities refused to allow Hahn to attend the December 10 award ceremony. His absence raised unwelcomed questions about where Hahn was, why the Allies were holding him, and what had become of Heisenberg and the other Uranium Club scientists.

CHAPTER SIXTY-FIVE

VICTORY

A TALE OF TWO CITIES, WAR DEPARTMENT NEWSREEL

At 10:58 the morning of August 9, the bomb was exploded above the city, and in the towering mushroom, Japan could read its doom. This was more than a routine bombing. It was the funeral pyre of an aggressor nation.

WORLD WAR II NEWSREEL

For the valley area of little more than three square miles, blast and fire reduced the industrial plants and surrounding buildings to blackened rubble. The cloud rose to sixty thousand feet in less than five minutes and there it stayed, a boiling turbulent mass expanding until it reached some two miles across.

Three days after the bombing of Hiroshima, the *BBC* broadcast a report of a second bomb targeting Nagasaki, Japan. The bomb triggered a chain reaction that consumed just 2.2 pounds of its 13.6 pounds of plutonium. The energy from those 2.2 pounds leveled a good part of the city with a blast even more powerful than that of Hiroshima, a blast equivalent to 168,000 conventional 500-pound bombs striking the same aiming point at the same precise moment.[1]

Nagasaki was the back-up target that day. The primary target, the arsenal in the city of Kokura, was hidden under a blanket of smoke and haze, so Major Charles Sweeney, pilot of the B-29 carrying the bomb, moved on to Nagasaki. There, the aiming point was also obscured by clouds. Running short on fuel, he feared he would have to return to base with the bomb still onboard. But then a few miles farther, the bombardier spotted a hole in the clouds and released his payload.

It exploded in the sky midway between two Mitsubishi Munitions plants. Ironically, it was those plants that manufactured the torpedoes used to attack Pearl Harbor. Japanese air-launched torpedoes were designed to sink to a depth of seventy feet before rising to attack depth. The water in Pearl Harbor was just forty feet deep. The torpedoes had to be redesigned, and Mitsubishi did the work. Their shallow-running torpedoes sank eight U.S. Navy ships, killing a large number of navy personnel among the more than two thousand servicemen, women, and civilians who lost their lives that day.[2]

The precise number of Japanese killed by the two bombs is unknown. Estimates range from 110,000 to 210,000 with Hiroshima deaths roughly twice those of Nagasaki.[3] Hopes ran high after Nagasaki that Oppenheimer's promise to his Los Alamos recruits—that their work would shorten the war and save American lives—would come true. But it took one more blow to convince the Japanese to surrender. The evening after the Nagasaki bombing when the captive scientists at Farm Hall tuned in to the 7:00 p.m. *BBC* broadcast, A-bombs were already old news. The lead story was not the Nagasaki bombing, it was "Soviets declare war on Japan; troops attack Manchuria." The Japanese and Soviets had signed a neutrality pact at the beginning of the war. The Japanese were counting on the Soviets to help them negotiate reasonable peace terms with the United States. Now, faced with the prospects of more atomic bomb attacks and a two-front war against the Soviets and the United States, Japan finally decided to surrender. Had they not surrendered, Groves expected to have three implosion bombs ready for use by September and possibly seven in December.[4]

Never had the invention of a new weapon changed the nature of warfare so dramatically. Gun powder had taken a century and the airplane, a generation. But the atom bomb changed the calculus of future conflicts overnight.[5] Optimists hoped it would make a world war among nuclear-armed adversaries too horrific to contemplate. Pessimists feared it would trigger a global arms race and bring the use of another atom bomb—intentional or accidental—closer to reality.

CHAPTER SIXTY-SIX

RADIATION POISONING REPORTED

CBS NEWS, AUGUST 14, 1945, SEVEN PM EASTERN WAR TIME, BOB TROUT REPORTING

"The Japanese have accepted our terms fully." That's the word we just received from the White House in Washington. This, ladies, and gentlemen, is the end of the Second World War.

"VJ DAY CELEBRATION IN TIMES SQUARE," *NBC*, AUGUST 14, 1945, BEN GRAUER REPORTING

This is the NBC mobile unit in the heart, center and the pulsing focus of a joyous nation at Times Square in New York City, Ben Grauer reporting as I stand, literally stand on tip toes sticking my head outside the silver-gray top of the NBC mobile unit and look uptown over a booming, seething, thriving mass of faces lifted joyously and buoyantly and gloriously happy at the final conclusion of a desperate war.

On September 2, 1945, six hundred journalists assembled onboard the battleship *Missouri* in Tokyo harbor to witness the Japanese surrender. As Allied and Japanese diplomats and military officers stepped forward to sign the documents, an overwhelming 85 percent of Americans approved of using the atomic bomb.[1] American prestige around the world could not have been higher. But while Americans and their European allies celebrated the war's end, Oppenheimer and Groves were dealing with some troubling news. Tokyo radio was reporting that many Hiroshima residents who had escaped injury from the bomb or incurred only minor scrapes and bruises were suddenly suffering from strange and often fatal illnesses. "Even those who received minor burns, and looked quite healthy at first," the radio reported, "weakened after a few days for some unknown reason."[2]

General Groves consulted medical experts who told him it was not likely that the illnesses were caused by radiation, so he and Oppenheimer dismissed the reports as Japanese propaganda. "We are not bothered a bit," he said, "excepting for what they are trying to do is create sympathy [for the Japanese]."[3] But then three days after the surrender ceremony, Wilfred Burchett, an Australian war correspondent, filed a story that could not be easily dismissed as Japanese propaganda. While his fellow journalists were covering the Japanese surrender, he struck a deal with Japan's Domei news agency. The agency's staffers hadn't heard from their Hiroshima correspondent Nakamura since the bombing, and Burchett was eager to get to Hiroshima. Burchett agreed to try to locate him, and Domei gave Burchett a letter of introduction to facilitate his efforts.[4]

After a twenty-one-hour train ride seated across from Japanese military officers glowering at Burchett, he made it to Hiroshima and managed to track down a translator and Nakamura. Together they made their way through a city that "looked like a monster steamroller had passed over it" to what was left of the city's main hospital. The hallways and wards were crammed with bomb victims. Most of the injuries—the lacerations, concussions, broken and severed limbs, and third- and fourth-degree burns—were of a kind that Burchett had seen too many times before caused by conventional explosive and incendiary bombs. But there were also a surprising number of survivors who had not been injured in the initial explosion and now, almost a month after the bombing, were developing symptoms of a mysterious illness. It began with diarrhea, nausea, chronic fatigue, and hair loss along with a diminishing white blood cell count. In the worst cases, black and red blotches developed on the skin, gums, ears, and nose and began to bleed. Temperatures spiked, flesh rotted, and the patients died.[5]

Burchett knew that military censors would not allow him to report the unusual, likely radiation-related illnesses, so he conspired with Japanese journalists to elude the censors and wire his eyewitness account to London where the *Daily Express* featured it on their front page:

30TH Day in Hiroshima: Those who escaped begin to die, victims of
THE ATOMIC PLAGUE
WILFRED BURCHETT: "I write this as a warning to the world"
DOCTORS FALL AS THEY WORK
Poison Gas fear: All wear masks[6]

"In Hiroshima, thirty days after the first atomic bomb destroyed the city and shook the world," Burchett wrote, "people are still dying, mysteriously and horribly—people who were uninjured in the cataclysm—from an unknown something which I can only describe as the atomic plague."[7]

Anticipating a torrent of press inquiries in the aftermath of the Hiroshima bombing, General Groves hired a *New York Times* science reporter, William Laurence, and gave him exclusive access to report on the Trinity test and the Hiroshima and Nagasaki bombing missions. He also commissioned physicist Henry DeWolf Smyth to write an official history of the development of the atom bomb. The articles and the history were published several weeks after the Japanese surrender. By lifting the cloak of secrecy, Groves hoped to shape the news coverage, but the accounts of a mysterious illness caught him off guard.

He told reporters that the bomb had been detonated several hundred meters above the city so that the radioactive fallout would be dispersed in the stratosphere. That was only partially true. The winds at that altitude did scatter much of the fallout. But the USAAF opted for an airburst because the bomb's blast would damage a larger area than a ground explosion. The reduction in radioactive fallout was a fortuitous, but unintended, side effect.[8]

Groves had no sympathy for the Japanese. They had brought the bombing of Hiroshima and Nagasaki on themselves with their attack on Pearl Harbor and the atrocities committed on American POWs in Bataan and in Japanese prisons. They deserved what they got. Still, he worried if the reports were believed, a bomb that continued to kill long after it exploded might raise thorny moral issues. The Geneva Convention signed after World War I had drawn a red line. It outlawed the use of poison gas and chemical and biological weapons in an armed conflict. General Groves and Oppenheimer worried that the atom bomb might be perceived as crossing that line.[9]

To counter the radiation disease stories, Groves sent his second-in-command, Brigadier General Thomas Farrell, and a team of radiation experts to Hiroshima and Nagasaki to investigate. He also pounced on a statement by a Harvard University physicist who told a reporter that Hiroshima would remain uninhabitable for seventy years. Quoting Oppenheimer, Groves called the claim lunacy, and the Harvard professor—after questioning by the FBI—issued a statement clarifying his comment.[10] At the White House, President Truman's press secretary sent Groves a memo suggesting that he prove the radiation posed no long-term danger by inviting a group of reporters to meet him and Oppenheimer at the Trinity site.

On September 9, two months after the Trinity test, the army opened the gates surrounding ground zero and admitted twenty newspapermen, photographers, and scientists. Radiologists carrying Geiger counters led the way as the group approached a shallow 1,200-foot-wide crater with a "green, glass-like coating resembling fine jade" where the bomb tower once stood. Unlike Hiroshima, the Trinity bomb had been detonated a mere hundred feet above the ground. As a result, much of the radioactive fallout was concentrated in the crater and had not yet fully decayed. Project scientists assured the reporters it would be safe "within a relatively short time."[11]

The tour had the effect that Groves, Oppenheimer, and the White House intended: further discrediting the claim from the Harvard physicist. A United Press correspondent wrote that the reporters left "convinced that they would not die within a few weeks or that their hair and teeth would not fall out as Jap propaganda would have the world believe."[12] The Associated Press assured readers that "the facts do not bear out Japanese propaganda."[13]

When General Farrell cabled the results of his investigation in Hiroshima, he too contradicted the Japanese claims. The radiation levels in the city were low and "could not have been detrimental to the health of persons . . . living in the bombed areas," he said.[14] An army surgeon reported similar findings from a survey in Nagasaki: "radioactivity in the Nagasaki area is virtually negligible and constitutes no hazard to health."[15]

The reports were accurate but also misleading. Groves's researchers were analyzing the wrong type of radiation. Fallout was made up primarily of radioactive debris. It was produced a minute after detonation and stayed radioactive for anywhere from a couple of minutes to years. The mysterious illnesses were caused by initial radiation, which was made up of high-energy gamma rays and neutrons generated in the first minute after detonation. It was short-lived, lasting less than a second, but in that time, it penetrated the victims' skin, damaging internal organs and killing off white blood cells, which the body needed to heal itself. The symptoms were not immediately apparent. But over a period of days to weeks, as more white blood cells died off, the victims became sicker and sicker. The greater the radiation exposure, the more serious the illness.[16]

During the war, Manhattan Project researchers had conducted extensive animal experiments to determine the biological effects of radiation and, according to University of California historian Sean Malloy, they had demonstrated "most of the serious radiation injuries the Japanese doctors had encountered . . . including dramatic destruction of white blood cells and fatal hemorrhagic diseases." Groves's policy of compartmentalization and his penchant for secrecy meant that "little or nothing about the Health Division findings" made their way to Los Alamos.[17] When they did, Oppenheimer "never really paid too much attention," said atom bomb historian Alex Wellerstein. And "because Oppenheimer didn't know/care about radiation effects, General Leslie Groves didn't really, either."[18] They may not have been worried about the effect on the Japanese, but they were concerned about the safety of the flight crew who would drop the bomb and the Allied forces who might occupy the bombed area.

Declassified War Department documents revealed that Oppenheimer sent General Farrell a cable before the Trinity test warning: "During the detonation, [gamma] radiations are emitted which (unless personnel is shielded) are expected to be injurious with a radius of a mile and lethal with a radius of about six-tenths of a mile."[19] And General Groves cabled his chief of staff after the test warning: "To persons who are completely unshielded, gamma rays may be lethal to 3,500 feet and neutrons to about 2,000 feet."[20]

Once the evidence of radiation-induced illnesses became undeniable, General Groves and Oppenheimer attempted to downplay the horror of radiation poisoning. In December 1945, General Groves took that effort to the halls of Congress, telling a Senate Committee the doctors had told him those who suffered radiation injury from the initial explosion died instantly "and those whose exposure was not enough to kill them instantly would die without undo suffering. In fact, they say it is a very pleasant way to die."[21] Oppenheimer and Groves "weren't shining heroes" nor "despicable villains," said historian Wellerstein. "These guys were not really in the 'cover-up' business. They certainly were in the 'spin' business."[22]

Six months after Groves testified, U.S. authorities revealed the specific source of the mysterious disease, gamma rays: "Fifteen to 20 per cent of the atomic bomb deaths in Japan were caused by the new killers of the atomic age—gamma and other lethal rays—the United States Strategic Bombing Survey disclosed today."[23]

CHAPTER SIXTY-SEVEN

TRINITY FALLOUT

Chemists at Eastman-Kodak in Rochester, New York, first noticed the problem in the autumn of 1945. And it kept getting worse. The fourteen-by-seventeen-inch film sheets they manufactured for industrial x-rays had imperfections. Tiny dots and lines were showing up on what should have been uniformly clean film. The company's chemists eventually traced the problem to contaminants in the cardboard used to separate the film sheets. But the cardboard had been manufactured in two different paper mills five hundred miles apart, one in Indiana and the other in Iowa. How could they both be contaminated? Searching for clues, Kodak's investigators homed in on the process used to produce the cardboard.

Both companies milled straw, mixed it with various materials, added water, and then formed the mix into cardboard sheets. Laboratory analysis revealed that the water, which came from several locations, was the source of the problem. It was contaminated with Cerium, a metal normally found in the western United States. How it got to Ohio and Indiana was not clear. Even more puzzling, the Cerium, which was normally stable and didn't emit radiation, was radioactive.

The mystery was eventually solved after news of the Trinity test in New Mexico became public. The explosion had emitted a burst of neutrons and, just like Fermi's neutron bombardment experiments in Rome, they had turned a stable, non-radioactive element Cerium into a radioactive element, Cerium 141. The radioactive soil mixed with the fallout from the explosion and was carried by the wind and river water more than 900 miles to the northeast where it entered the Wabash and Iowa river watersheds that supplied water to the two paper mills. The mills used that water to make the cardboard which was then shipped to Kodak in Rochester, New York.[1]

The x-ray contamination was the least of Groves' and Oppenheimer's fallout problems. In the rush to test the plutonium bomb, Manhattan Project scientists had predicted that the fallout would be dangerous in a 15-mile radius of ground zero. In fact, the danger zone extended closer to 150 miles and included areas downwind from Trinity that were mistakenly believed to be uninhabited.[2] Unaware of the Trinity test or the radiation danger, local farmers and ranchers struggled to make sense of several bizarre phenomena: a strange white ash rained from the sky staining linens hanging on clotheslines and leaving a snow-like coating on crops and rooftops.[3] Dairy cows lost their hair and, when it grew back, its color had changed from brown to white.[4] A black cat turned white, and one half of a rancher's beard turned gray.[5]

The people of New Mexico had not been warned in 1945 about the atomic test. When an unusually large number of babies died that year, no one thought to tie the deaths to radiation poisoning. It wasn't until decades later that researchers discovered that the New Mexico infant mortality rate spiked from 89.1 deaths per thousand live births in 1944 to 100.8 deaths in 1945, the year of the Trinity test, and declined the year after to 78.2 per thousand.[6] The precise number of New Mexico residents who died because of the Trinity test will never be known.[7] Years later when "Downwinders," as they were called, organized and petitioned Washington for compensation, their petition was denied.[8]

Downwinders near the Hanford nuclear plants suffering from cancers and other diseases fared only marginally better. Hanford had continued producing plutonium after the war to counter the Soviets who successfully tested their first atom bomb in August 1949. In lawsuits that dragged on for twenty-five years an estimated 3500 to 5000 plaintiffs suffering from various cancers, autoimmune, and other diseases sued the Department of Energy. The vast majority of those suits were dismissed because they couldn't meet the scientific threshold required to prove their diseases were caused by radiation . . . or the victims died. Hanford workers, on the other hand, whose radiation exposure had been closely monitored, received up to $150,000 if they met the necessary criteria.[9]

EPILOGUE

Samuel Goudsmit left Michigan after the war for a teaching position at Northwestern University and then moved on in 1948 to Brookhaven National Laboratory in New York where he was a senior scientist and Chairman of the Physics Department.

In 1947, he published *Alsos*, his World War II memoir on the Alsos mission. The memoir drew from classified documents, including the Farm Hall transcripts, which would not become public for almost a half century, so it helped shape the views of journalists and historians all during the post-war era. In it he cited reasons he believed Germany's atom bomb program had failed. The attacks on the Norsk Hydro heavy water plant were obviously critical. But beyond that, he zeroed in on Nazi dogma: the hiring of party loyalists for important managerial positions, exiling of Jewish scientists, bureaucratic infighting, and the blind hero worship of the program's chief architect, Werner Heisenberg.[1]

"Werner Heisenberg . . . was the foremost atomic physicist in Germany," Goudsmit wrote. "No good young German scientist would think of questioning the word of the master." Not similarly constrained, and still bitter about his parents' murder, Goudsmit saved much of his criticism for his former friend, arguing that Heisenberg had only a superficial understanding of how a reactor or an atom bomb works.[2] Goudsmit's memoir triggered an exchange of angry letters between the two men. In the end, after Heisenberg produced German war records, Goudsmit conceded that some of his criticism had been inaccurate and that he had underestimated Heisenberg's knowledge. The two ultimately agreed to a truce.[3]

Werner Heisenberg returned to a war-ravaged Germany and the university town of Göttingen where he had met Fermi, and Oppenheimer. He held several prestigious positions, all the while trying to restore Germany's theoretical physics program to world prominence. Those who knew him described him as often depressed. Never again did his star burn as brightly as it had before the war.[4]

In 1947, Heisenberg was invited to the United States to give a series of lectures. Victor Weiskopf, a former Manhattan Project physicist who had done post-doctoral work with Heisenberg in Göttingen before the war and was now on the faculty at MIT, decided it was time to put Heisenberg's wartime record in the past. Accordingly, he arranged a reception for Heisenberg at his home and asked several top scientists from MIT, Harvard, and elsewhere to attend. Only about half of them accepted the invitation.

A decade later, Heisenberg's colleagues tried to rewrite the past, telling Robert Jungk, author of the controversial book *Brighter Than a Thousand Suns*, that they never wanted to build Hitler an atom bomb. Heisenberg refused to be interviewed for the book, and he never claimed to have sabotaged the atom bomb program.

Postwar accounts tended to place Germany's premier theoretical physicist at one extreme or the other: either a saboteur, and therefore a traitor to the fatherland, or a second-rate nuclear engineer.[5] Nothing Heisenberg said could change those narratives, so he eventually stopped trying. For the final decades of his life, he remained silent about the war.

Robert Oppenheimer returned to academia. For almost twenty years he served as director of the Institute for Advanced Study in Princeton, New Jersey. Showered with accolades, he became a national hero, appearing on the covers of *Time* and *Life* magazines, and an advisor to presidents. In late 1945, as the full horror of Hiroshima and Nagasaki began to sink in, Oppenheimer met President Truman for the first time. Normally self-assured and articulate, Oppenheimer mumbled and seemed hesitant. When the president asked if something was wrong, he confessed, "I feel I have blood on my hands." Incensed, the commander in chief issued a stern rebuke. "The blood is on my hands. Let me worry about that."[6]

In the 1950s, with the nuclear arms race taking off, Oppenheimer became a leading advocate for arms control. His opposition to building a hydrogen bomb and a bitter feud with a commissioner on the Atomic Energy Commission won him some formidable enemies. In 1954, at the height of the McCarthy red scare era, the AEC struck a humiliating blow. Dredging up Oppenheimer's prewar

Communist ties, they voted to strip him of his top-secret security clearance. After that, Oppenheimer declined to get involved in public debate over the Eisenhower administration's nuclear policy and instead became an intellectual outsider lecturing on the role of science in the modern world.

In 1963, President Lyndon Johnson awarded Oppenheimer the prestigious Enrico Fermi prize for public service, an act widely interpreted as political rehabilitation. Oppenheimer resigned his position as director of Princeton's Institute of Advanced Study in 1966, and a year later died of throat cancer. Fifty-five years after his death, Department of Energy secretary Jennifer Granholm ruled that the AEC decision to revoke Oppenheimer's security clearance was the result of a flawed process that was biased and unfair and reinstated his security clearance.

Enrico Fermi went on to have a distinguished career at the University of Chicago and to work the job he loved the most, teaching physics. Six of his students from Chicago and one from Rome became Nobel Prize winners themselves.[7] When not in the classroom, he was in the laboratory, conducting research and making important contributions in particle physics and other fields.

Fermi initially opposed development of the hydrogen bomb, which was far more powerful than the Nagasaki bomb, but then reversed himself during the Korean War when the Soviet Union began building one. To his critics who opposed building the bombs, Fermi offered no apologies. "It is no good trying to stop knowledge from going forward," he said. "Whatever nature has in store for mankind, unpleasant as it may be, men must accept, for ignorance is never better than knowledge."[8]

Fermi died of stomach cancer in 1954 at the age of fifty-three. But his legacy continues to inspire. University of Chicago students study nuclear science in the Enrico Fermi Institute. In Washington, D.C., the U.S. Department of Energy administers the Enrico Fermi Award to encourage excellence in energy science research benefitting mankind. The DOE also honored his legacy by naming the Fermi National Accelerator Laboratory (Fermilab) and the Fermi Gamma Ray Space Telescope after him.

Lieutenant General Leslie Groves, celebrated by the *Saturday Evening Post* as "The Atom General" and by *Collier's* magazine as "The Man Who Made Manhattan," spent several months traveling the country making speeches, accepting awards, and honoring those who helped make the Manhattan Project a success, while also winding down the project's operations.

But behind closed doors, scientists whom Groves had alienated during the war and government officials were organizing and asserting themselves on critical issues—most importantly, who would control atomic energy: civilians or the military. When Congress voted to create the civilian-run Atomic Energy Commission (AEC), Groves criticized it for putting a cumbersome government bureaucracy in charge of decision-making and excluding the military. David Lilienthal, the chairman of the AEC, took umbrage and, according to Groves, "wanted no advice of any kind from me. He wanted nothing whatsoever to do with me."[9] Lilienthal wasn't the only one Groves alienated.

On January 30, 1948, General Groves met with General Dwight Eisenhower, the army chief of staff, for his annual performance review. The two men never particularly liked each other, and Eisenhower came prepared with a long list of criticisms. Near the top was Groves's treatment of fellow officers, which Eisenhower considered "ruthless to the point of cruelty," "arrogant and insensitive." Next on the list, his callous treatment of subordinates. But it was his shameless pursuit of advancement when named to head the Manhattan Project that rankled Eisenhower the most. He had maneuvered past those in front of him in the line of promotion from colonel to brigadier general in 1942, and then to temporary major general in 1944, and finally to lieutenant general in 1948 when he "used his influence with members of Congress to blackmail the army leadership" into promoting him.[10]

His military career stalled, and his reputation tarnished, fifty-one-year-old Leslie Groves decided to retire. After twenty-nine years in the army, he took a job as vice president in charge of research and development at Remington Rand and remained in the public eye, making speeches, and commenting on atom bomb–related issues until he retired to private life in 1961 at the age of sixty-five.[11]

ACKNOWLEDGMENTS

This book would not have been possible without the help of the authors (all listed in my bibliography) whose research helped inform and shape *The Greatest Scientific Gamble.*

I owe a very particular debt to Richard Rhodes, whose authoritative *The Making of the Atomic Bomb* stirred my interest in the bomb's history and who generously agreed to be interviewed. Thanks, too, to Brian Schwartz, professor of physics, Brooklyn College and the Graduate Center, City University of New York, who helped school me in the fine points of nuclear fission and greatly expanded my understanding of the engineering challenges involved in building the atom bomb.

Cindy Kelly, the founder and president of the Atomic Heritage Foundation, was my go-to source when I needed a contact in Los Alamos or the Department of Energy. Her videotaped interviews with Manhattan Project veterans and historians posted on the foundation's website served as a valuable resource. Olivia Fermi humanized her grandfather for me by sharing personal anecdotes about him acclimating to life in the United States. Another valuable resource: the librarians at the Emilio Segrè Visual Archives of the American Institute of Physics and its digitized collection of Samuel Goudsmit's papers, which facilitated my research.

I am grateful for the support of friends, colleagues, and family, among them Tim Smith, Jeffrey Stark, David Wald, Tom Seligson, and my brother Gordon, who read drafts of the book and offered invaluable suggestions. Special thanks to my friend and colleague John Merrow, a constant source of encouragement and advice, especially during my search for a publisher.

The Greatest Scientific Gamble is my second book about the former friends-turned-rivals. The first was a short e-book published by Amazon titled *Chasing Heisenberg: The Race for the Atom Bomb*. I'm indebted to my friend Sara

Nelson, senior vice president and executive editor at Harper Collins, for her encouragement and support launching that book. She put me in touch with her friend Andrew Rosenheim, a novelist and Rhodes Scholar, who was my editor on *Chasing Heisenberg* and chief advocate at Amazon as I navigated the various prepublication hurdles. His encouragement led me to revise and expand that e-book. After he left Amazon to work on his own projects, he generously continued to offer counsel and support.

Finally, I want to thank my computer-savvy children, Matthew and Karen, who came to my rescue when my MacBook Pro acted up, and my wife and fellow writer, Joanne Kaufman, who read several drafts and never stopped reminding me that most readers are in trouble most of the time—so explain, explain, explain all that science!

NOTES

CHAPTER ONE. A SMALL GLOBAL COMMUNITY

1. Thomas Powers, *Heisenberg's War: The Secret History of the German Bomb* (Da Capo Press, 2000), 10.
2. David C. Cassidy, *Beyond Uncertainty: Heisenberg, Quantum Physics, and the Bomb* (Bellevue Literary Press, 2009), 363.
3. David C. Cassidy, *Uncertainty: The Life and Science of Werner Heisenberg* (W. H. Freeman, 1992), 303.
4. Repository AIP.org, Samuel A. Goudsmit papers, Box 46, Folder 148; Werner Heisenberg, Yearbook of the American Philosophical Society, 1976, 77.
5. Daniel Lang, "A Farewell to String and Sealing Wax," in *From Hiroshima to the Moon: Chronicles of Life in the Atomic Age* (Simon & Schuster, 1959), 221.
6. Werner Heisenberg, *Physics and Beyond: Encounters and Conversations*, trans. Arnold J. Pomerans (Harper & Row, 1971), 165.
7. Lang, "Farewell to String and Sealing Wax," 218.
8. Martijn van Calmthout, *Sam Goudsmit and the Hunt for Hitler's Atom Bomb*, trans. Michiel Horn (Prometheus Books, 2018), 38.
9. Ibid., 40.
10. Laura Fermi, *Atoms in the Family: My Life with Enrico Fermi* (University of Chicago Press, 1954), 32.
11. "Tinkering with Angry Atoms May Blow Up the Earth," *The Sun*, September 28, 1919; "Nature's Greatest Secret, the Might of the Atom, Stirs Scientists," *New York Herald*, September 28, 1919.
12. H. G. Wells, *The World Set Free* (Cabin John, MD: Wildside Press LLC), 83.

CHAPTER TWO. A MODERN ALCHEMIST

1. "This Month in Physics History, May 1932: Chadwick Reports the Discovery of the Neutron," *APS News* 16, no. 5 (2007).
2. Atoms with the same number of protons in their nucleus are called elements. All atoms with thirteen protons in their nucleus, for example, are the element aluminum. Atoms with fourteen protons in their nucleus are the element silicon. In 1934, there were eighty-six known elements.
3. Laura Fermi, *Atoms in the Family: My Life with Enrico Fermi* (University of Chicago Press, 1954), 89.
4. Stephane Groueff, *Manhattan Project: The Untold Story of the Making of the Atomic Bomb* (iUniverse, 2000), 25.
5. Isaac Asimov, *Atom: Journey across the Subatomic Cosmos* (Truman Talley Books/Plume, 1992), 167.
6. Fermi, *Atoms in the Family*, 87.
7. Ibid., 86.
8. Richard Rhodes, *The Making of the Atomic Bomb* (Simon & Schuster, 1986), 210–211.
9. Ibid., 211, 209.
10. Brian Schwartz, email to author, August 13, 2021.
11. Rhodes, *The Making of the Atomic Bomb*, 213.
12. Klaus Hoffmann, *Otto Hahn: Achievement and Responsibility* (Springer-Verlag, 1993), 116.
13. Fermi, *Atoms in the Family*, 91–92.
14. Rhodes, *The Making of the Atomic Bomb*, 228.
15. Ibid., 217.
16. Bruce Cameron Reed, *The History and Science of the Manhattan Project*, 2nd ed. (Springer, 2019), 67.
17. Fermi, *Atoms in the Family*, 98.
18. Brian Schwartz, email to author, August 13, 2021.

CHAPTER THREE. NOBEL PRIZE

1. Laura Fermi, *Atoms in the Family: My Life with Enrico Fermi* (University of Chicago Press, 1954), 115.
2. Ibid., 118.

3. "Nobel Prize in Physics 1938," The Nobel Prize, Nobelprize.org/prizes/physics/1938/summary.
4. Fermi, *Atoms in the Family*, 123.
5. David N. Schwartz, *The Last Man Who Knew Everything: The Life and Times of Enrico Fermi, Father of the Nuclear Age* (Basic Books, 2017), 141.
6. Ibid., 141.
7. Ibid., 142.
8. Ibid., 134.
9. Richard Rhodes, *The Making of the Atomic Bomb* (Simon & Schuster, 1986), 230.

CHAPTER FOUR. NUCLEAR FISSION

1. Richard Rhodes, *The Making of the Atomic Bomb* (Simon & Schuster, 1986), 235.
2. Klaus Hoffmann, *Otto Hahn: Achievement and Responsibility* (Springer-Verlag, 1993), 118, 120.
3. David Bodanis, *E=MC²: A Biography of the World's Most Famous Equation* (Berkeley Books, 2001), 106.
4. Robert Jungk, *Brighter Than a Thousand Suns: A Personal History of the Atomic Scientists*, trans. James Cleugh (Harcourt, Brace, 1958), 62.
5. Hoffmann, *Otto Hahn*, 136.
6. Ibid., 141.
7. Walter Isaacson, *Einstein: His Life and Universe* (Simon & Schuster, 2007), 139.

CHAPTER FIVE. NEWS OF NUCLEAR FISSION ROCKS SCIENTIFIC WORLD

1. Enrico Fermi Nobel Lecture, December 12, 1938, 417.
2. "6 Elements Found in Uranium Atom," *New York Times*, February 25, 1939.
3. Richard Rhodes, interview with the author, November 21, 2014.
4. "Vast Energy Freed by Uranium Atom," *New York Times*, January 31, 1939.
5. "Emilio Segrè's Interview," *Voices of the Manhattan Project*, The Atomic Heritage Foundation, https://ahf.nuclearmuseum.org/voices/oral-histories/emilio-segres-interview/.

6. Richard Rhodes, *The Making of the Atomic Bomb* (Simon & Schuster, 1986), 296.
7. Gino Segrè and Bettina Hoerlin, *The Pope of Physics: Enrico Fermi and the Birth of the Atomic Age* (Henry Holt and Company, 2016), 151, 152.
8. Ibid., 302.

CHAPTER SIX. BOHR'S DISCOVERY

1. Richard Rhodes, *The Making of the Atomic Bomb* (Simon & Schuster, 1986), 284.
2. Waldemar Kaempffert, "The Story Behind the Atom Bomb," *New York Times*, August 12, 1945, 63.
3. Rhodes, *The Making of the Atomic Bomb*, 294.
4. Winston S. Churchill, *The Second World War: The Gathering Storm* (Houghton Mifflin, 1976), 345.
5. Jean Medawar and David Pyke, *Hitler's Gift: Scientists Who Fled Nazi Germany* (Richard Cohen Books, 2000), 235.

CHAPTER SEVEN. HEISENBERG IN MICHIGAN

1. David C. Cassidy, *Uncertainty: The Life and Science of Werner Heisenberg* (W. H. Freeman, 1992), 390.
2. David Irving, *The Virus House: Germany's Atomic Research and Allied Counter-Measures* (William Kimber, 1967), 142.
3. Elisabeth Heisenberg, *Inner Exile: Recollections of a Life with Werner Heisenberg*, trans. S. Cappellari and C. Morris (Birkhäuser, 1984), 57.
4. Arnold Kramish, *The Griffin: The Greatest Untold Espionage Story of World War II* (Houghton Mifflin, 1986), 44.
5. Cassidy, *Uncertainty*, 264–265.
6. Thomas Powers, *Heisenberg's War: The Secret History of the German Bomb* (Da Capo Press, 2000), 4.
7. Heisenberg, *Inner Exile*, 64.
8. David C. Cassidy, *J. Robert Oppenheimer and the American Century* (Pi Press, 2005), 167.
9. Repository AIP.org, Samuel A. Goudsmit papers, Box 46, Folder 148, Werner Heisenberg, Yearbook of the American Philosophical Society, 1976, 77.
10. Walter Isaacson, *Einstein: His Life and Universe* (Simon & Schuster, 2007), 84.

11. "J. Robert Oppenheimer," Atomicarchive.com.
12. Paul Lawrence Rose, *Heisenberg and the Nazi Atomic Bomb Project: A Study in German Culture* (University of California Press, 1998), 317.
13. "Today's Events in Pictures, Scientists Investigate Cyclotron Research at Purdue," *Indianapolis Star*, July 7, 1939, 15.
14. Gino Segrè and Bettina Hoerlin, *The Pope of Physics: Enrico Fermi and the Birth of the Atomic Age* (Henry Holt and Company, 2016), 162.
15. Werner Heisenberg, *Physics and Beyond: Encounters and Conversations*, trans. Arnold J. Pomerans (Harper & Row, 1971), 169.
16. Ibid., 170.
17. Daniel Lang, "A Farewell to String and Sealing Wax," in *From Hiroshima to the Moon: Chronicles of Life in the Atomic Age* (Simon & Schuster, 1959), 241.
18. Kramish, *Griffin*, 117.
19. Cassidy, *Uncertainty*, 303.
20. Samuel A. Goudsmit, *Alsos* (American Institute of Physics Press, 1996), 115.
21. Ibid., 114.
22. David C. Cassidy, *Beyond Uncertainty: Heisenberg, Quantum Physics, and the Bomb* (Bellevue Literary Press, 2009), 296.

CHAPTER EIGHT. LIGHTNING WARFARE

1. Tadeusz Piotrowski, *Poland's Holocaust: Ethnic Strife, Collaboration with Occupying Forces and Genocide in the Second Republic 1918–1947* (McFarland & Co), 301.
2. Arnold Kramish, *The Griffin: The Greatest Untold Espionage Story of World War II* (Houghton Mifflin, 1986), 59.
3. Neal Bascomb, *The Winter Fortress: The Epic Mission to Sabotage Hitler's Atomic Bomb* (Houghton Mifflin Harcourt, 2016), 11.
4. David Irving, *The Virus House: Germany's Atomic Research and Allied Counter-Measures* (William Kimber, 1967), 43.
5. Bascomb, *Winter Fortress*, 12.
6. Richard von Schirach, *The Night of the Physicists: Operation Epsilon: Heisenberg, Hahn, Weizsäcker and the German Bomb*, trans. Simon Pare (Haus Publishing, 2014), 25.
7. Irving, *Virus House*, 59.
8. Jeremy Bernstein, *Hitler's Uranium Club: The Secret Recordings at Farm Hall*, 2nd ed. (Copernicus Books, 2001), xxv.

9. Irving, *Virus House*, 49.
10. Jim Baggott, *The First War of Physics: The Secret History of the Atom Bomb 1939–1949* (Pegasus Books, 2010), 38.
11. David C. Cassidy, *Uncertainty: The Life and Science of Werner Heisenberg* (W. H. Freeman, 1992), 426, 425, 428.
12. Ibid., 427.
13. Von Schirach, *Night of Physicists*, 147.
14. Kramish, *Griffin*, 53.
15. Damien Lewis, *Hunting the Nazi Bomb: The Special Forces Mission to Sabotage Hitler's Deadliest Weapon* (Quercus Editions Ltd, 2017), 75.
16. David Bodanis, *E= MC²: A Biography of the World's Most Famous Equation* (Berkeley Books, 2001), 129.
17. William L. Laurence, "Vast Power Source in Atomic Energy Opened by Science," *New York Times*, May 5, 1940, 51.
18. Sam Kean, *The Bastard Brigade: The True Story of the Renegade Scientists and Spies Who Sabotaged the Nazi Atomic Bomb* (Little, Brown and Company 2019), 121.
19. Thomas Powers, *Heisenberg's War: The Secret History of the German Bomb* (Da Capo Press, 2000), 82.
20. Bernstein, *Hitler's Uranium Club*, 48.
21. Jim Baggott, *The First War of Physics: The Secret History of the Atom Bomb 1939–1949* (Pegasus Books, 2010), 60–61.
22. The committee's name was chosen arbitrarily as a code name. It comes from Maud Ray Kent, a governess working for a famous Danish physicist.
23. "MAUD Committee Report," The Atomic Heritage Foundation, https://ahf.nuclearmuseum.org/ahf/key-documents/maud-committee-report/.

CHAPTER NINE. A SECOND ROUTE TO AN ATOMIC BOMB

1. Andrew J. Rotter, *Hiroshima: The World's Bomb* (Oxford University Press, 2008), 90.
2. Richard Rhodes, *The Making of the Atomic Bomb* (Simon & Schuster, 1986), 350–351.
3. Ibid., 351.
4. Thomas Powers, *Heisenberg's War: The Secret History of the German Bomb* (Da Capo Press, 2000), 97.
5. Ibid., 108.
6. Robert Jungk, *Brighter Than a Thousand Suns: A Personal History of the Atomic Scientists,*

trans. James Cleugh (Harcourt, Brace, 1958), 96–97.

7. Powers, *Heisenberg's War*, 100.

CHAPTER TEN. THE BATTLE OF BRITAIN

1. Richard Hallion, *The Accuracy of Air-Delivered Weapons* (Pathfinder, Air Power Centre Bulletin, October 2019).
2. Alfred Price, *Instruments of Darkness: The History of Electronic Warfare* (Charles Scribner's Sons, 1978), 38–39.
3. Brian Johnson, *The Secret War* (Pen & Sword Military Classics, 2004), 19–20, 17–19.
4. R. V. Jones, *Most Secret War* (Wordsworth, 1998), 89.
5. Price, *Instruments of Darkness*, 27.
6. Ibid., 27, 39.
7. Peter Stansky, *The First Day of the Blitz: September 7, 1940* (Yale University Press, 2007), 3.
8. David C. Cassidy, *Uncertainty: The Life and Science of Werner Heisenberg* (W. H. Freeman, 1992), 427.
9. Daniel Lang, "A Farewell to String and Sealing Wax," in *From Hiroshima to the Moon: Chronicles of Life in the Atomic Age* (Simon & Schuster, 1959), 224.

CHAPTER ELEVEN. ENRICO AND LAURA IN AMERICA

1. Laura Fermi, *Atoms in the Family: My Life with Enrico Fermi* (University of Chicago Press, 1954), 143.
2. "To Fermi with Love, Part 3," The Atomic Heritage Foundation, https://ahf.nuclearmuseum.org/.
3. Andrew Nagorski, *Hinterland: American Eyewitness to the Nazi Rise to Power* (Simon & Schuster, 2012), 206.
4. "Dictatorship Said Planned by Terrorists: 17 Men Arrested; More Sought in Plot to Kill Congressmen, Jews," *AP Redding Record*, January 15, 1940.
5. "German-Americans Stand under Swastika in N.Y. Denounce 'International Jewry,'" *Austin Daily Herald*, September 21, 1939.
6. Fermi, *Atoms in the Family*, 170.
7. Ibid., 171.

CHAPTER TWELVE. HARNESSING ATOMIC ENERGY

1. Laura Fermi, *Atoms in the Family: My Life with Enrico Fermi* (University of Chicago Press, 1954), 171.
2. Richard Rhodes, *The Making of the Atomic Bomb* (Simon & Schuster, 1986), 394.
3. Brian Schwartz, email to author, August 13, 2021.
4. Rhodes, *The Making of the Atomic Bomb*, 345.
5. Sam Kean, *The Bastard Brigade: The True Story of the Renegade Scientists and Spies Who Sabotaged the Nazi Atomic Bomb* (Little, Brown and Company 2019); Werner Heisenberg, *Physics and Beyond: Encounters and Conversations*, trans. Arnold J. Pomerans (Harper & Row, 1971), 180.
6. Arnold Kramish, *The Griffin: The Greatest Untold Espionage Story of World War II* (Houghton Mifflin, 1986), 175.
7. Ibid., p. 396.
8. David Irving, *The Virus House: Germany's Atomic Research and Allied Counter-Measures* (William Kimber, 1967), 114.

CHAPTER THIRTEEN. HEAVY WATER

1. Neal Bascomb, *The Winter Fortress: The Epic Mission to Sabotage Hitler's Atomic Bomb* (Houghton Mifflin Harcourt, 2016), 5.
2. Ibid., 4.
3. Ibid., 15.
4. Thomas Powers, *Heisenberg's War: The Secret History of the German Bomb* (Da Capo Press, 2000), 195.
5. Daniel Lang, "A Farewell to String and Sealing Wax," in *From Hiroshima to the Moon: Chronicles of Life in the Atomic Age* (Simon & Schuster, 1959), 226.
6. David Irving, *The Virus House: Germany's Atomic Research and Allied Counter-Measures* (William Kimber, 1967), 157.

CHAPTER FOURTEEN. LIFE IN LEIPZIG AND BERLIN

1. David C. Cassidy, *Uncertainty: The Life and Science of Werner Heisenberg* (W. H. Freeman, 1992), 429, 430.

2. David Irving, *The Virus House: Germany's Atomic Research and Allied Counter-Measures* (William Kimber, 1967), 88, 89.
3. Ibid., 434.

CHAPTER FIFTEEN. THE MANHATTAN PROJECT

1. Gino Segrè and Bettina Hoerlin, *The Pope of Physics: Enrico Fermi and the Birth of the Atomic Age* (Henry Holt and Company, 2016), 178, 175, 180.
2. Ibid., 177.
3. Glenn T. Seaborg and Eric Seaborg, *Adventures in the Atomic Age: From Watts to Washington* (Farrar, Straus and Giroux, 2001), 86.
4. Ibid., 86–87.
5. Michael Gannon, *Operation Drumbeat: The Dramatic True Story of Germany's First U-Boat Attacks Along the American Coast in World War II* (Harper & Row Publishers, 1990), 388.
6. "Werner Heisenberg Biographical," The Nobel Prize, https://www.nobelprize.org/prizes/physics/1932/heisenberg/biographical.
7. Rainer Karlsch and Mark Walker, "New Light on Hitler's Bomb," *Physicsworld* (June 1, 2005).
8. David Irving, *The Virus House: Germany's Atomic Research and Allied Counter-Measures* (William Kimber, 1967), 172.
9. David C. Cassidy, *Uncertainty: The Life and Science of Werner Heisenberg* (W. H. Freeman, 1992), 427.
10. Irving, *Virus House*, 166.
11. David C. Cassidy, *Beyond Uncertainty: Heisenberg, Quantum Physics, and the Bomb* (Bellevue Literary Press, 2009), 333.
12. Irving, *Virus House*, 166.
13. Ibid., 167.
14. John Cornwell, *Hitler's Scientists: Science, War and the Devil's Pact* (Penguin Books, 2003), 320.
15. Cassidy, *Uncertainty*, 152.
16. Michael Eckert, "Werner Heisenberg: Controversial Scientist," *Physics World* (November 30, 2001), 4–5.
17. Irving, *Virus House*, 167.
18. Ibid., 168.

CHAPTER SIXTEEN. A MEETING OF WEAPONS DEVELOPERS

1. Jim Baggott, *The First War of Physics: The Secret History of the Atom Bomb 1939–1949* (Pegasus Books, 2010), 111.
2. Thomas Powers, *Heisenberg's War: The Secret History of the German Bomb* (Da Capo Press, 2000), 133.
3. Baggott, *First War of Physics*, 126.
4. Sam Kean, *The Bastard Brigade: The True Story of the Renegade Scientists and Spies Who Sabotaged the Nazi Atomic Bomb* (Little, Brown and Company, 2019), 137.
5. Ibid., 126–127.
6. Powers, *Heisenberg's War*, 137.
7. Elisabeth Heisenberg, *Inner Exile: Recollections of a Life with Werner Heisenberg*, trans. S. Cappellari and C. Morris (Birkhäuser, 1984), 74.
8. David Irving, *The Virus House: Germany's Atomic Research and Allied Counter-Measures* (William Kimber, 1967), 135.
9. Heisenberg, *Inner Exile*, 74.
10. Irving, *Virus House*, 134.
11. Jeremy Bernstein, *Hitler's Uranium Club: The Secret Recordings at Farm Hall*, 2nd ed. (Copernicus Books, 2001), xxviii.
12. Albert Speer, *Inside the Third Reich: Memoirs*, trans. Richard and Clara Winston (Simon & Schuster, 1970), 227.
13. Powers, *Heisenberg's War*, 145–146.
14. Speer, *Inside the Third Reich*, 227.
15. Irving, *Virus House*, 138.

CHAPTER SEVENTEEN. GENERAL LESLIE GROVES

1. "Appointment of the Military Policy Committee," The George C. Marshal Foundation Library, September 23, 1942.
2. Robert S. Norris, *Racing for the Bomb: General Leslie R. Groves, the Manhattan Project's Indispensable Man* (Steerforth Press, 2002), 181.
3. Richard Rhodes, *The Making of the Atomic Bomb* (Simon & Schuster, 1986), 426.
4. Norris, *Racing for the Bomb*, 175.
5. Ibid., 174.

6. Stephane Groueff, *Manhattan Project: The Untold Story of the Making of the Atomic Bomb* (iUniverse, 2000), 8.
7. Ibid., 176.
8. Leslie R. Groves, *Now It Can Be Told: The Story of the Manhattan Project* (Da Capo Press, 1983), 5.
9. Norris, *Racing for the Bomb*, 607.
10. Groves, *Now It Can Be Told*, 28.
11. William Lawren, *The General and the Bomb: A Biography of General Leslie R. Groves, Director of the Manhattan Project* (Dodd Mead & Company, 1988), 38.
12. Ibid., 40.
13. "Atomic Mission," in *The Corps of Engineers: Construction in the United States*, by Lenore Fine and Jesse A. Remington (Center of Military History, U.S. Army, 1989), 663.
14. Lawren, *The General and the Bomb*, 69–70.
15. Ibid., 69–70.
16. William J. Wilcox, "K-25 Plant: Forty-four Acres and a Mile Long," in *The Manhattan Project: The Birth of the Atomic Bomb in the Words of Its Creators, Eyewitnesses, and Historians*, ed. Cynthia Kelly (Black Dog & Leventhal Publishers, 2007), 199.
17. Ibid., 199.
18. Groueff, *Manhattan Project*, 98.
19. Ibid., 22.
20. Norris, *Racing for the Bomb*, 208.
21. Groves, *Now It Can Be Told*, 103.
22. Lawren, *The General and the Bomb*, 232.
23. Groves, *Now It Can Be Told*, 40.
24. Ibid., 40.
25. Lawren, *The General and the Bomb*, 80.
26. Groueff, *Manhattan Project*, 35.
27. Ibid., 35.
28. "William J. Wilcox Interview," Voices of the Manhattan Project, The Atomic Heritage Foundation, ahf.nuclearmuseum.org.
29. Groueff, *Manhattan Project*, 39.
30. Groves, *Now It Can Be Told*, 19, 72.
31. "Atomic Mission," in *The Corps of Engineers: Construction in the United States*, by Lenore Fine and Jesse A. Remington (Center of Military History, U.S. Army, 1989), 668.
32. Lawren, *The General and the Bomb*, 91.
33. Ibid., 92.

34. Groueff, *Manhattan Project*, 101.
35. Ibid., 105.
36. Ibid., 103.
37. Lawren, *The General and the Bomb*, 92, 105.

CHAPTER EIGHTEEN. GENERAL GROVES MEETS ROBERT OPPENHEIMER

1. Richard Rhodes, *The Making of the Atomic Bomb* (Simon & Schuster, 1986), 449.
2. William Lawren, *The General and the Bomb: A Biography of General Leslie R. Groves, Director of the Manhattan Project* (Dodd Mead & Company, 1988), 100.
3. Gregg Herken, *Brotherhood of the Bomb: The Tangled Lives and Loyalties of Robert Oppenheimer, Ernest Lawrence, and Edward Teller* (Henry Holt and Company, 2002), 55.
4. Jim Baggott, *The First War of Physics: The Secret History of the Atom Bomb 1939–1949* (Pegasus Books, 2010), 144.

CHAPTER NINETEEN. OPERATION FRESHMAN

1. David Irving, *The Virus House: Germany's Atomic Research and Allied Counter-Measures* (William Kimber, 1967), 124.
2. Neal Bascomb, *The Winter Fortress: The Epic Mission to Sabotage Hitler's Atomic Bomb* (Houghton Mifflin Harcourt, 2016), ix, 95.
3. Ibid., 108.
4. Ibid., 158–162.
5. Robert S. Norris, *Racing for the Bomb: General Leslie R. Groves, the Manhattan Project's Indispensable Man* (Steerforth Press, 2002), 283.
6. Bascomb, *Winter Fortress*, 310.
7. David C. Cassidy, *Beyond Uncertainty: Heisenberg, Quantum Physics, and the Bomb* (Bellevue Literary Press, 2009), 333.
8. Mark Walker, *Nazi Science: Myth, Truth, and the German Atomic Bomb* (Perseus Publishing, 1995), 141.
9. Arnold Kramish, *The Griffin: The Greatest Untold Espionage Story of World War II* (Houghton Mifflin, 1986), 118; Cassidy, *Beyond Uncertainty*, 341.
10. Walker, *Nazi Science*, 167–168.

11. Kramish, *Griffin*, 118.

CHAPTER TWENTY. CHICAGO

1. Richard Rhodes, *The Making of the Atomic Bomb* (Simon & Schuster, 1986), 432.
2. "Wise Gets Confirmation: Checks with State Department on Nazis' Extermination Campaign," *New York Times*, November 25, 1942.
3. Rhodes, *The Making of the Atomic Bomb*, 433–436.
4. Gino Segrè and Bettina Hoerlin, *The Pope of Physics: Enrico Fermi and the Birth of the Atomic Age* (Henry Holt and Company, 2016), 186, 434.
5. Jim Baggott, *The First War of Physics: The Secret History of the Atom Bomb 1939–1949* (Pegasus Books, 2010), 131.
6. Klaus Hoffmann, *Otto Hahn: Achievement and Responsibility* (Springer-Verlag, 1993), 165.
7. Arnold Kramish, *The Griffin: The Greatest Untold Espionage Story of World War II* (Houghton Mifflin, 1986), 164.
8. Baggott, *First War of Physics*, 260.
9. "Herbert Anderson Interview," *Voices of the Manhattan Project*, The Atomic Heritage Foundation, https://ahf.nuclearmuseum.org/voices/oral-histories/herbert-andersons-interview-1965/.
10. Rhodes, *The Making of the Atomic Bomb*, 439.
11. Stephane Groueff, *Manhattan Project: The Untold Story of the Making of the Atomic Bomb* (iUniverse, 2000), 88.
12. Rhodes, *The Making of the Atomic Bomb*, 442.
13. Ibid., 440–442.
14. Ibid., 436.
15. Kramish, *Griffin*, 63.

CHAPTER TWENTY-ONE. SS *MONT-BLANC*

1. Richard Rhodes, *The Making of the Atomic Bomb* (Simon & Schuster, 1986), 674.
2. Jim Baggott, *The First War of Physics: The Secret History of the Atom Bomb 1939–1949* (Pegasus Books, 2010), 95.
3. Ibid., 95.
4. Toronto Daily Star, December 8, 1917.

5. Laura M. MacDonald, *Curse of The Narrows: The Halifax Disaster of 1917* (Walker Publishing Company, 2005), 288.

CHAPTER TWENTY-TWO. CONSTRUCTION

1. "Statement by the President Announcing the Use of the A-Bomb at Hiroshima," Harry S. Truman Library & Museum, August 6, 1945.
2. John Coster-Mullen, *Atom Bombs: The Top Secret Inside Story of Little Boy and Fat Man* (self-pub., 2009), 41.
3. Vincent C. Jones, *Manhattan: The Army and the Atomic Bomb* (Center of Military History, U.S. Army, 1985), 132.
4. William Lawren, *The General and the Bomb: A Biography of General Leslie R. Groves, Director of the Manhattan Project* (Dodd Mead & Company, 1988), 154.
5. Stephane Groueff, *Manhattan Project: The Untold Story of the Making of the Atomic Bomb* (iUniverse, 2000), 166.
6. Robert S. Norris, *Racing for the Bomb: General Leslie R. Groves, the Manhattan Project's Indispensable Man* (Steerforth Press, 2002), 218, 218–221.
7. Daniel Lang, "The New Yorker, A Reporter at Large, Career at Y-12," in *From Hiroshima to the Moon* (Simon & Shuster, 1959), 26.
8. Michelle Gerber, "Termination Winds," in *The Manhattan Project: The Birth of the Atomic Bomb in the Words of Its Creators, Eyewitnesses, and Historians*, ed. Cynthia Kelly (Black Dog & Leventhal Publishers, 2007), 190.
9. S. L. Sanger and Craig Wollner, *Working on the Bomb: An Oral History of WWII Hanford* (Continuing Education Press, Portland State University, 1995), 142.
10. Richard Rhodes, *The Making of the Atomic Bomb* (Simon & Schuster, 1986), 499.
11. Kai Bird and Martin J. Sherwin, *American Prometheus: The Triumph and Tragedy of J. Robert Oppenheimer* (Vintage Books, 2005), 215.
12. William Lawren, *The General and the Bomb: A Biography of General Leslie R. Groves, Director of the Manhattan Project* (Dodd Mead & Company, 1988), 127.
13. Ibid., 128.
14. Daniel Lang, "A Farewell to String and Sealing Wax," in *From Hiroshima to the Moon: Chronicles of Life in the Atomic Age* (Simon & Schuster, 1959), 224, 225.

CHAPTER TWENTY-THREE. HEAVY WATER II

1. David Irving, *The Virus House: Germany's Atomic Research and Allied Counter-Measures* (William Kimber, 1967), 156.
2. Ibid., 183–184.
3. Barton J. Bernstein, "The Uneasy Alliance: Roosevelt, Churchill, and the Atomic Bomb, 1940–1945." *Western Political Quarterly* (June 1976): 206–208.
4. Robert S. Norris, *Racing for the Bomb: General Leslie R. Groves, the Manhattan Project's Indispensable Man* (Steerforth Press, 2002), 343.
5. Arnold Kramish, *The Griffin: The Greatest Untold Espionage Story of World War II* (Houghton Mifflin, 1986), 164.

CHAPTER TWENTY-FOUR. LOS ALAMOS

1. David Bodanis, *$E=MC^2$: A Biography of the World's Most Famous Equation* (Berkeley Books, 2001), 147.
2. Ibid., 145.
3. Robert Serber, *The Los Alamos Primer: The First Lectures on How to Build an Atomic Bomb* (University of California Press, 1992), xi.
4. Richard Rhodes, *The Making of the Atomic Bomb* (Simon & Schuster, 1986), 462.
5. Ibid., 199.
6. Ibid., 323.
7. Ibid., 323.
8. Kai Bird and Martin J. Sherwin, *American Prometheus: The Triumph and Tragedy of J. Robert Oppenheimer* (Vintage Books, 2005), 209.
9. Ibid., 232.
10. Ibid., 231.
11. Ibid., 43, 49.
12. Ibid., 60.
13. Rhodes, *The Making of the Atomic Bomb*, 570.
14. Ibid., 148.
15. Bird and Sherwin, *American Prometheus*, 209.
16. Ibid., 216.
17. Ibid., 216.

CHAPTER TWENTY-FIVE. NOTHING BUT THEORY

1. Robert S. Norris, *Racing for the Bomb: General Leslie R. Groves, the Manhattan Project's Indispensable Man* (Steerforth Press, 2002), 239.
2. Lillian Hoddeson, Paul W. Henriksen, Roger A. Meade, and Catherine Westfall, *Critical Assembly: A Technical History of Los Alamos during the Oppenheimer Years, 1943–1945* (Cambridge University Press, 2004), 77.
3. Richard Rhodes, *The Making of the Atomic Bomb* (Simon & Schuster, 1986), 465
4. Leslie R. Groves, *Now It Can Be Told: The Story of the Manhattan Project* (Da Capo Press, 1983), 72.
5. William Lawren, *The General and the Bomb: A Biography of General Leslie R. Groves, Director of the Manhattan Project* (Dodd Mead & Company, 1988), 168.
6. Samuel A. Goudsmit, *Alsos* (American Institute of Physics Press, 1996), 65.
7. Ferenc Morton Szasz, *The Day the Sun Rose Twice: The Story of the Trinity Site Nuclear Explosion, July 16, 1945* (University of New Mexico Press, 1984), 24.

CHAPTER TWENTY-SIX. DUMB LUCK

1. Stephane Groueff, *Manhattan Project: The Untold Story of the Making of the Atomic Bomb* (iUniverse, 2000), 91.
2. Ibid., 124–125.
3. Ibid., 127.

CHAPTER TWENTY-SEVEN. HUNTING HEISENBERG

1. Werner Heisenberg, *Physics and Beyond: Encounters and Conversations*, trans. Arnold J. Pomerans (Harper & Row, 1971), 188.
2. John Cornwell, *Hitler's Scientists: Science, War and the Devil's Pact* (Penguin Books, 2003), 321.
3. Thomas Powers, *Heisenberg's War: The Secret History of the German Bomb* (Da Capo Press, 2000), 203.
4. Damien Lewis, *Hunting the Nazi Bomb: The Special Forces Mission to Sabotage Hitler's Deadliest Weapon* (Quercus Editions Ltd, 2017), 289.
5. Powers, *Heisenberg's War*, 204.

6. Leslie R. Groves, *Now It Can Be Told: The Story of the Manhattan Project* (Da Capo Press, 1983), 199.
7. Robert S. Norris, *Racing for the Bomb: General Leslie R. Groves, the Manhattan Project's Indispensable Man* (Steerforth Press, 2002), 293.
8. Groves, *Now It Can Be Told*, 216.

CHAPTER TWENTY-EIGHT. PEENEMÜNDE

1. David Irving, *The Virus House: Germany's Atomic Research and Allied Counter-Measures* (William Kimber, 1967), 216.
2. Ibid., 216.
3. Cruewell had been captured as a result of a colossal blunder. During an aerial inspection tour of German troops in North Africa, his pilot mistook British troops for friendly Italians. He landed and he and his passenger were taken prisoner.
4. Brian Johnson, *The Secret War* (Pen & Sword Military Classics, 2004), 256–257.
5. Mark Murphy, *The Exploits of Agent 110: Allen Dulles in Wartime* (CIA Studies in Intelligence), CIA.gov, 69.
6. Arnold Kramish, *The Griffin: The Greatest Untold Espionage Story of World War II* (Houghton Mifflin, 1986), 101.
7. R. V. Jones, *Most Secret War* (Wordsworth, 1998), 339.
8. Murphy, *The Exploits of Agent 110*, 69.
9. Jones, *Most Secret War*, 340.
10. Ibid., 346.

CHAPTER TWENTY-NINE. HAMBURG

1. Malcolm Gladwell, *The Bomber Mafia: A Dream, a Temptation, and the Longest Night of the Second World War* (Little, Brown and Company, 2021), 110.
2. Richard Overy, *The Bombers and the Bombed: Allied Air War Over Europe, 1940–1945* (Penguin, 2013), 144.
3. Richard Rhodes, *The Making of the Atomic Bomb* (Simon & Schuster, 1986), 476.
4. Overy, *The Bombers and the Bombed*, 136.
5. "Red Sky Over Hamburg," *New York Times*, July 26, 1943, 18.
6. Roger Daniels, *Franklin D. Roosevelt: The War Years, 1939–1945* (University of Illinois

Press 2016), 519.

7. Thomas Powers, *Heisenberg's War: The Secret History of the German Bomb* (Da Capo Press, 2000), 203–204.
8. Jennet Conant, *Man of the Hour: James B. Conant, Warrior Scientist* (Simon & Schuster, 2017), 301.
9. David Irving, *The Virus House: Germany's Atomic Research and Allied Counter-Measures* (William Kimber, 1967), 212.
10. Vincent C. Jones, *Manhattan: The Army and the Atomic Bomb* (Center of Military History, U.S. Army, 1985), 283.
11. Robert S. Norris, *Racing for the Bomb: General Leslie R. Groves, the Manhattan Project's Indispensable Man* (Steerforth Press, 2002), 297.
12. Leslie R. Groves, *Now It Can Be Told: The Story of the Manhattan Project* (Da Capo Press, 1983), 206.

CHAPTER THIRTY. THE KELLEX COMPANY IS HIRING

1. Stephane Groueff, *Manhattan Project: The Untold Story of the Making of the Atomic Bomb* (iUniverse, 2000), 184.
2. Ibid., 185–186.

CHAPTER THIRTY-ONE. ALSOS

1. When his intelligence unit suggested the code name Alsos, General Groves didn't know it meant "groves" in Greek. Code names were supposed to be innocuous so that they didn't draw undue attention to the work. He was infuriated when he found out but decided that changing the name would draw even more attention, so he left it.
2. Martijn van Calmthout, *Sam Goudsmit and the Hunt for Hitler's Atom Bomb*, trans. Michiel Horn (Prometheus Books, 2018), 128.
3. Jeremy Bernstein, *Hitler's Uranium Club: The Secret Recordings at Farm Hall*, 2nd ed. (Copernicus Books, 2001), 44.
4. Jeffrey T. Richelson, *Spying on the Bomb: American Nuclear Intelligence from Nazi Germany to Iran and North Korea* (W. W. Norton & Company, 2006), 29.
5. Ibid., 32.
6. Memorandum for the Secretary of War, Atomic Fission Bombs, from L. R. Groves, Major

General, U.S.A., National Security Archive, April 23, 1945.

7. Richelson, *Spying on the Bomb*, 32.
8. Thomas Powers, *Heisenberg's War: The Secret History of the German Bomb* (Da Capo Press, 2000), 285.
9. Richelson, *Spying on the Bomb*, 33.
10. Ibid., 4.

CHAPTER THIRTY-TWO. HEAVY WATER III

1. Arnold Kramish, *The Griffin: The Greatest Untold Espionage Story of World War II* (Houghton Mifflin, 1986), 170.
2. Robert S. Norris, *Racing for the Bomb: General Leslie R. Groves, the Manhattan Project's Indispensable Man* (Steerforth Press, 2002), 284.
3. Thomas Powers, *Heisenberg's War: The Secret History of the German Bomb* (Da Capo Press, 2000), 212.
4. Jim Baggott, *The First War of Physics: The Secret History of the Atom Bomb 1939–1949* (Pegasus Books, 2010), 216.
5. Tami Davis Biddle, *Rhetoric and Reality in Air Warfare: The Evolution of British and American Ideas about Strat3egic Bombing, 1914–1945* (Princeton University Press, 2002), 223.
6. Neal Bascomb, *The Winter Fortress: The Epic Mission to Sabotage Hitler's Atomic Bomb* (Houghton Mifflin Harcourt, 2016), 269.
7. Richard von Schirach, *The Night of the Physicists: Operation Epsilon: Heisenberg, Hahn, Weizsäcker and the German Bomb*, trans. Simon Pare (Haus Publishing, 2014), 141.
8. David Irving, *The Virus House: Germany's Atomic Research and Allied Counter-Measures* (William Kimber, 1967), 221.
9. Jeremy Bernstein, *Hitler's Uranium Club: The Secret Recordings at Farm Hall*, 2nd ed. (Copernicus Books, 2001), 35.

CHAPTER THIRTY-THREE. "NICHOLAS BAKER" AT LOS ALAMOS

1. Elisabeth Heisenberg, *Inner Exile: Recollections of a Life with Werner Heisenberg*, trans. S. Cappellari and C. Morris (Birkhäuser, 1984), 77.

2. Letter from Heisenberg to Bohr, Niels Bohr Scientific Correspondence—Supplement Collection, Niels Bohr Archives, September 14, 1939.
3. Amir D. Aczel, *Uranium Wars: The Scientific Rivalry That Created the Nuclear Age* (Palgrave MacMillan, 2009), 147.
4. Thomas Powers, *Heisenberg's War: The Secret History of the German Bomb* (Da Capo Press, 2000), 255.
5. In his 1986 book *The Griffin: The Greatest Untold Espionage Story of World War II* (Houghton Mifflin, 1986), Arnold Kramish, a Manhattan Project physicist and author, wrote that the Ministry of Education, which oversaw Germany's atomic research program at the time, requested information about the U.S. atom bomb program. Heisenberg's colleague C. F. Weiszäcker suggested that Niels Bohr might be a good source and arranged for Heisenberg to be invited to lecture in Copenhagen (118–119).
6. Powers, *Heisenberg's War*, 253.
7. Ibid., 257.

CHAPTER THIRTY-FOUR. GLENN SEABORG

1. Glenn T. Seaborg and Eric Seaborg, *Adventures in the Atomic Age: From Watts to Washington* (Farrar, Straus and Giroux, 2001), 105, 106.
2. Ibid., 97.
3. Ibid., 112.
4. Claudia Clark, *Radium Girls: Women and Industrial Health Reform, 1910–1935* (University of North Carolina Press, 1997), introduction.
5. John McPhee, *The Curve of Binding Energy: A Journey into the Awesome and Alarming World of Theodore B. Taylor* (Farrar, Straus and Giroux, 1977), 44.

CHAPTER THIRTY-FIVE. MANUFACTURING PROBLEMS

1. Vincent C. Jones, *Manhattan: The Army and the Atomic Bomb* (Center of Military History, U.S. Army, 1985), 155.
2. Ibid., 157.
3. Stephane Groueff, *Manhattan Project: The Untold Story of the Making of the Atomic Bomb* (iUniverse, 2000), 313.

CHAPTER THIRTY-SIX. NORSK HYDRO SHIPS HEAVY WATER

1. Damien Lewis, *Hunting the Nazi Bomb: The Special Forces Mission to Sabotage Hitler's Deadliest Weapon* (Quercus Editions Ltd, 2017), 373.
2. Ibid., 241–242.
3. Neal Bascomb, *The Winter Fortress: The Epic Mission to Sabotage Hitler's Atomic Bomb* (Houghton Mifflin Harcourt, 2016), 60.
4. Richard Rhodes, *The Making of the Atomic Bomb* (Simon & Schuster, 1986), 514.
5. David Irving, *The Virus House: Germany's Atomic Research and Allied Counter-Measures* (William Kimber, 1967), 243.
6. Ibid., 241.
7. Bascomb, *The Winter Fortress*, 306.

CHAPTER THIRTY-SEVEN. OSS KIDNAP PLOT

1. Robert S. Norris, *Racing for the Bomb: General Leslie R. Groves, the Manhattan Project's Indispensable Man* (Steerforth Press, 2002), 291–292.
2. Thomas Powers, *Heisenberg's War: The Secret History of the German Bomb* (Da Capo Press, 2000), 267.
3. Ibid., 265.
4. Ibid., 265.
5. Ibid., 264, 267.
6. Ibid., 313.

CHAPTER THIRTY-EIGHT. UNDERGROUND BUNKER

1. David C. Cassidy, *Uncertainty: The Life and Science of Werner Heisenberg* (W. H. Freeman, 1992), 487.
2. Ibid., 487.
3. Ibid., 486, 488.
4. Jim Baggott, *The First War of Physics: The Secret History of the Atom Bomb 1939–1949* (Pegasus Books, 2010), 270.
5. Cassidy, *Uncertainty*, 488.

6. Ibid., 489.
7. Richard von Schirach, *The Night of the Physicists: Operation Epsilon: Heisenberg, Hahn, Weizsäcker and the German Bomb*, trans. Simon Pare (Haus Publishing, 2014), 137.

CHAPTER THIRTY-NINE. PLUTONIUM PROBLEM

1. Bruce Cameron Reed, *The History and Science of the Manhattan Project*, 2nd ed. (Springer, 2019), 327.
2. Robert S. Norris, *Racing for the Bomb: General Leslie R. Groves, the Manhattan Project's Indispensable Man* (Steerforth Press, 2002), 362.
3. S. L. Sanger and Craig Wollner, *Working on the Bomb: An Oral History of WWII Hanford* (Continuing Education Press, Portland State University, 1995), 3.
4. Norris, *Racing for the Bomb*, 221.
5. "Atomic Mission," in *The Corps of Engineers: Construction in the United States*, by Lenore Fine and Jesse A. Remington (Center of Military History, U.S. Army, 1989), 661.
6. Norris, *Racing for the Bomb*, 364.

CHAPTER FORTY. A SECRET MEETING

1. Richard Rhodes, *The Making of the Atomic Bomb* (Simon & Schuster, 1986), 552.
2. Ibid., 550.
3. Ibid., 552.
4. Jim Baggott, *The First War of Physics: The Secret History of the Atom Bomb 1939–1949* (Pegasus Books, 2010), 280.
5. Robert S. Norris, *Racing for the Bomb: General Leslie R. Groves, the Manhattan Project's Indispensable Man* (Steerforth Press, 2002), 367.
6. Stephane Groueff, *Manhattan Project: The Untold Story of the Making of the Atomic Bomb* (iUniverse, 2000), 317.
7. Ibid., 318–319.

CHAPTER FORTY-ONE. WUNDERWAFFEN

1. Arnold Kramish, *The Griffin: The Greatest Untold Espionage Story of World War II* (Houghton Mifflin, 1986), 155.

CHAPTER FORTY-TWO. DARKEST HOURS

1. Kai Bird and Martin J. Sherwin, *American Prometheus: The Triumph and Tragedy of J. Robert Oppenheimer* (Vintage Books, 2005), 281.
2. Ibid., 280–281.
3. William Lawren, *The General and the Bomb: A Biography of General Leslie R. Groves, Director of the Manhattan Project* (Dodd Mead & Company, 1988), 260.
4. Richard Rhodes, *The Making of the Atomic Bomb* (Simon & Schuster, 1986), 467.
5. John McPhee, *The Curve of Binding Energy: A Journey into the Awesome and Alarming World of Theodore B. Taylor* (Farrar, Straus and Giroux, 1977), 81.
6. David N. Schwartz, *The Last Man Who Knew Everything: The Life and Times of Enrico Fermi, Father of the Nuclear Age* (Basic Books, 2017), 237.
7. Jennet Conant, *Man of the Hour: James B. Conant, Warrior Scientist* (Simon & Schuster, 2017), 310.
8. Conant, *Man of the Hour*, 310.
9. Laura Fermi, *Atoms in the Family: My Life with Enrico Fermi* (University of Chicago Press, 1954), 209, 211.
10. Stephane Groueff, *Manhattan Project: The Untold Story of the Making of the Atomic Bomb* (iUniverse, 2000), 328.
11. Ibid., 320.
12. Rhodes, *The Making of the Atomic Bomb*, 543.
13. Richard G. Hewlett and Oscar E. Anderson Jr., *A History of the United States Atomic Energy Commission*, vol. 1: *The New World 1939/1946* (Pennsylvania State University Press, 1962), 375.

CHAPTER FORTY-THREE. ALSOS IN PARIS

1. Leslie R. Groves, *Now It Can Be Told: The Story of the Manhattan Project* (Da Capo Press, 1983), 193.

2. Daniel Lang, "A Farewell to String and Sealing Wax," in *From Hiroshima to the Moon: Chronicles of Life in the Atomic Age* (Simon & Schuster, 1959), 232.
3. American Institute of Physics, Niels Bohr Archives, Samuel Goudsmit Papers, Box 25, Folder 03, Alsos—travel and other orders and receipts with some Goudsmit correspondence, 1942–1945.
4. Jim Baggott, *The First War of Physics: The Secret History of the Atom Bomb 1939–1949* (Pegasus Books, 2010), 274.
5. Samuel A. Goudsmit, *Alsos* (American Institute of Physics Press, 1996), 56.
6. Ibid., 54.
7. Ibid., 57.
8. Ibid., 60.
9. David Irving, *The Virus House: Germany's Atomic Research and Allied Counter-Measures* (William Kimber, 1967), 300.
10. Goudsmit, *Alsos*, 65.
11. Irving, *Virus House*, 293.
12. Lang, "Farewell to String," 230–231.

CHAPTER FORTY-FOUR. HANFORD

1. Laura Fermi, *Atoms in the Family: My Life with Enrico Fermi* (University of Chicago Press, 1954), 200.
2. Gino Segrè and Bettina Hoerlin, *The Pope of Physics: Enrico Fermi and the Birth of the Atomic Age* (Henry Holt and Company, 2016), 212.
3. S. L. Sanger and Craig Wollner, *Working on the Bomb: An Oral History of WWII Hanford* (Continuing Education Press, Portland State University, 1995), 68.
4. Richard G. Hewlett and Oscar E. Anderson, Jr, *A History of the United States Atomic Energy Commission*, vol. 1: *The New World 1939/1946*, 220.
5. Vincent C. Jones, *Manhattan: The Army and the Atomic Bomb* (Center of Military History, U.S. Army, 1985), 215.
6. Stephen L. Sanger, "Working on the Bomb," in *The Manhattan Project*, ed. Cynthia Kelly (Black Dog & Leventhal Publishers, Inc., 2007), 193.
7. Richard Rhodes, *The Making of the Atomic Bomb* (Simon & Schuster, 1986), 498. David N. Schwartz, *The Last Man Who Knew Everything* (Basic Books Hachette Book Group, 2017), 228.
8. Rhodes, *The Making of The Atomic Bomb*, 557–558.

9. Stephane Groueff, *Manhattan Project: The Untold Story of the Making of the Atomic Bomb* (iUniverse, 2000), 302–303.
10. Ibid., 304.
11. Ibid., 304.
12. Ibid., 306.
13. Rhodes, *The Making of the Atomic Bomb*, 33.
14. Ibid., 558–560.
15. Neal Bascomb, *The Winter Fortress: The Epic Mission to Sabotage Hitler's Atomic Bomb* (Houghton Mifflin Harcourt, 2016), 311.
16. "Nazis Have Atom Bomb, Stockholm Report Declares," *Minneapolis Star*, October 3, 1944.
17. "Nazis' New 'V-2' Bomb a Failure According to Report of Witness," *Ottawa Journal*, October 5, 1944, 2.
18. Friedrich Georg, *Hitler's Miracle Weapons*, vol. 2: *The Secret History of the Rockets and Flying Craft of the Third Reich, from the V-1 to the A-9 Unconventional Short- and Medium-Ranged Weapons,* trans. C. F. Colton MA (Helion & Company, 2005), 50–51.

CHAPTER FORTY-FIVE. STRASBOURG LIBERATED

1. Samuel A. Goudsmit, *Alsos* (American Institute of Physics Press, 1996), 67.
2. American Institute of Physics, Niels Bohr Archives, Letter to Wife, Sub Series A, ALSOS Mission Material, Box 25, Folder 05, December 10, 1944.
3. David Irving, *The Virus House: Germany's Atomic Research and Allied Counter-Measures* (William Kimber, 1967), 307.
4. Sam Kean, *The Bastard Brigade: The True Story of the Renegade Scientists and Spies Who Sabotaged the Nazi Atomic Bomb* (Little, Brown and Company, 2019), 361.
5. Ibid., 307.
6. Jeremy Bernstein, *Hitler's Uranium Club: The Secret Recordings at Farm Hall,* 2nd ed. (Copernicus Books, 2001), 36.
7. "Reginal C. Augustine Interview," *Voices of the Manhattan Project*, The Atomic Heritage Foundation, https://ahf.nuclearmuseum.org/voices/oral-histories/reginald-c-augustines-interview/.
8. Thomas Powers, *Heisenberg's War: The Secret History of the German Bomb* (Da Capo Press, 2000), 371.
9. Kean, *Bastard Brigade*, 364.

10. Goudsmit, *Alsos*, 75.
11. American Institute of Physics, Niels Bohr Archives, Letter to Wife, Sub Series A, ALSOS Mission Material, Box 25, Folder 05, December 10, 1944.
12. Powers, *Heisenberg's War*, 371.
13. Goudsmit, *Alsos*, 69.
14. Ibid., 70.
15. Ibid., 71.
16. Ibid., 70.
17. Kean, *Bastard Brigade*, 362.

CHAPTER FORTY-SIX. RESTRICTED AREA

1. Jeffrey T. Richelson, *Spying on the Bomb: American Nuclear Intelligence from Nazi Germany to Iran and North Korea* (W. W. Norton & Company, 2006), 50.
2. Leslie R. Groves, *Now It Can Be Told: The Story of the Manhattan Project* (Da Capo Press, 1983), 218.
3. Thomas Powers, *Heisenberg's War: The Secret History of the German Bomb* (Da Capo Press, 2000), 376.
4. Richelson, *Spying on the Bomb*, 50.
5. Powers, *Heisenberg's War*, 377.

CHAPTER FORTY-SEVEN. BATTLE OF THE BULGE

1. "Germans Use Atomic Bomb," *Fort Worth Star-Telegram*, December 27, 1944.
2. Thomas Powers, *Heisenberg's War: The Secret History of the German Bomb* (Da Capo Press, 2000), 397–398, 392.
3. Ibid., 398.
4. Ibid., 402.
5. Jim Baggott, *The First War of Physics: The Secret History of the Atom Bomb 1939–1949* (Pegasus Books, 2010), 278.
6. Elisabeth Heisenberg, *Inner Exile: Recollections of a Life with Werner Heisenberg*, trans. S. Cappellari and C. Morris (Birkhäuser, 1984), 3.
7. Powers, *Heisenberg's War*, 403.

8. David C. Cassidy, *Uncertainty: The Life and Science of Werner Heisenberg* (W. H. Freeman, 1992), 493.

CHAPTER FORTY-EIGHT. BATTLE OF THE BULGE AFTERSHOCKS

1. Jim Schoke, email to author, August 20, 2015.
2. Robert S. Norris, *Racing for the Bomb: General Leslie R. Groves, the Manhattan Project's Indispensable Man* (Steerforth Press, 2002), 368.
3. Ibid., 369.
4. William Lawren, *The General and the Bomb: A Biography of General Leslie R. Groves, Director of the Manhattan Project* (Dodd Mead & Company, 1988), 203.
5. Ibid.,147.
6. Leslie R. Groves, *Now It Can Be Told: The Story of the Manhattan Project* (Da Capo Press, 1983), 31.

CHAPTER FORTY-NINE. THE LAST EXPERIMENT

1. David Irving, *The Virus House: Germany's Atomic Research and Allied Counter-Measures* (William Kimber, 1967), 314.
2. Ibid., 313.
3. Jim Baggott, *The First War of Physics: The Secret History of the Atom Bomb 1939–1949* (Pegasus Books, 2010), 288.
4. Irving, *Virus House,* 316.
5. Paul Lawrence Rose, *Heisenberg and the Nazi Atomic Bomb Project: A Study in German Culture* (University of California Press, 1998), 317.
6. Ibid., 30.
7. Richard von Schirach, *The Night of the Physicists: Operation Epsilon: Heisenberg, Hahn, Weizsäcker and the German Bomb,* trans. Simon Pare (Haus Publishing, 2014), 132.
8. Ibid.

CHAPTER FIFTY. DIEBNER'S LAB CAPTURED

1. Samuel A. Goudsmit, *Alsos* (American Institute of Physics Press, 1996), 88.
2. Ibid., 88.
3. Boris T. Pash, *The Alsos Mission: The Incredible Story of the Espionage Commandos Who Stole Hitler's Most Lethal Weapon* (Award Books, 1970), 197.

CHAPTER FIFTY-ONE. HEISENBERG'S FINAL EXPERIMENT

1. David Irving, *The Virus House: Germany's Atomic Research and Allied Counter-Measures* (William Kimber, 1967), 318–319.
2. Jim Baggott, *The First War of Physics: The Secret History of the Atom Bomb 1939–1949* (Pegasus Books, 2010), 289.
3. Samuel A. Goudsmit, *Alsos* (American Institute of Physics Press, 1996), xii.

CHAPTER FIFTY-TWO. THE GRIFFIN

1. Jim Baggott, *The First War of Physics: The Secret History of the Atom Bomb 1939–1949* (Pegasus Books, 2010), 322.
2. Arnold Kramish, *The Griffin: The Greatest Untold Espionage Story of World War II* (Houghton Mifflin, 1986), 117.
3. Owen Boycott, "Spy Left Out in the Cold How MI6 Buried heroic Exploits of Agent 'Griffin,'" *Guardian*, September 21, 2007.
4. Kramish, *Griffin*, 12, 25.
5. Ibid., 98.
6. Owen Boycott, "Spy Left Out in the Cold: How MI6 Buried heroic Exploits of Agent 'Griffin,'" *Guardian*, September 21, 2007.
7. Kramish, *Griffin*, 153.
8. Owen Boycott, "Spy Left out in the Cold: How MI6 Buried heroic Exploits of Agent 'Griffin,'" *Guardian*, September 21, 2007.

CHAPTER FIFTY-THREE. HUNTING HEISENBERG II

1. David C. Cassidy, *Uncertainty: The Life and Science of Werner Heisenberg* (W. H. Freeman, 1992), 497.
2. David Irving, *The Virus House: Germany's Atomic Research and Allied Counter-Measures* (William Kimber, 1967), 326, 332.
3. Geoffrey Brooks, *Hitler's Nuclear Weapons: The Development and Attempted Deployment of Radiological Armaments by Nazi Germany* (Pen & Sword Books, 1992), 134.
4. Samuel A. Goudsmit, *Alsos* (American Institute of Physics Press, 1996), 77–78.
5. Leslie R. Groves, *Now It Can Be Told: The Story of the Manhattan Project* (Da Capo Press, 1983), 231.
6. Martijn van Calmthout, *Sam Goudsmit and the Hunt for Hitler's Atom Bomb*, trans. Michiel Horn (Prometheus Books, 2018), 112–113.
7. Ibid., 113.
8. Boris T. Pash, *The Alsos Mission: The Incredible Story of the Espionage Commandos Who Stole Hitler's Most Lethal Weapon* (Award Books, 1970), 206.
9. Ibid., 207, 334.
10. Calmthout and Boekerij, *Goudsmit and the Hunt*, 118.
11. Goudsmit, *Alsos*, 99.
12. Thomas Powers, *Heisenberg's War: The Secret History of the German Bomb* (Da Capo Press, 2000), 424.

CHAPTER FIFTY-FOUR. CRITICAL MASS

1. Richard Rhodes, *The Making of the Atomic Bomb* (Simon & Schuster, 1986), 611.
2. Ibid., 610.
3. Ibid., 610.
4. Ibid., 611.
5. Ibid., 612.
6. William Lawren, *The General and the Bomb: A Biography of General Leslie R. Groves, Director of the Manhattan Project* (Dodd Mead & Company, 1988), 200.

CHAPTER FIFTY-FIVE. A CALL FROM THE SECRETARY OF WAR

1. Robert Norris, "Unprecedented Security Measures," in *The Manhattan Project: The Birth of the Atomic Bomb in the Words of Its Creators, Eyewitnesses, and Historians*, ed. Cynthia Kelly (Black Dog & Leventhal Publishers, 2007), 233.
2. Stephane Groueff, *Manhattan Project: The Untold Story of the Making of the Atomic Bomb* (iUniverse, 2000), 333.
3. Ibid., 333.
4. Daniel Lang, "The Atomic City," in *From Hiroshima to the Moon: Chronicles of Life in the Atomic Age* (Simon & Schuster, 1959), 22, 20.
5. Groueff, *Manhattan Project*, 334, 102.
6. Robert S. Norris, *Racing for the Bomb: General Leslie R. Groves, the Manhattan Project's Indispensable Man* (Steerforth Press, 2002), 208.
7. William J. Wilcox, "K-25 Plant: Forty-four Acres and a Mile Long," in *The Manhattan Project: The Birth of the Atomic Bomb in the Words of Its Creators, Eyewitnesses, and Historians*, ed. Cynthia Kelly (Black Dog & Leventhal Publishers, 2007), 199.
8. Groueff, *Manhattan Project*, 334.
9. With most of Tennessee's draft age men serving in the army or waiting to be called up, virtually all of the workers were female. Lang, "The Atomic City," in *From Hiroshima to the Moon*, 19–20.
10. Groueff, *Manhattan Project*, 336.
11. Richard Rhodes, *The Making of the Atomic Bomb* (Simon & Schuster, 1986), 602.
12. Groueff, *Manhattan Project*, 346.

CHAPTER FIFTY-SIX. HEISENBERG ON THE RUN

1. Jim Baggott, *The First War of Physics: The Secret History of the Atom Bomb 1939–1949* (Pegasus Books, 2010), 294.
2. Richard von Schirach, *The Night of the Physicists: Operation Epsilon: Heisenberg, Hahn, Weizsäcker and the German Bomb*, trans. Simon Pare (Haus Publishing, 2014), 9.
3. David C. Cassidy, *Uncertainty: The Life and Science of Werner Heisenberg* (W. H. Freeman, 1992), 499.
4. Boris T. Pash, *The Alsos Mission: The Incredible Story of the Espionage Commandos Who Stole Hitler's Most Lethal Weapon* (Award Books, 1970), 232.

5. Ibid., 239.
6. Cassidy, *Uncertainty*, 498.
7. Werner Heisenberg, *Physics and Beyond: Encounters and Conversations*, trans. Arnold J. Pomerans (Harper & Row, 1971), 191.
8. Ibid., 393.

CHAPTER FIFTY-SEVEN. THE HUNT IS OVER

1. Martijn van Calmthout, *Sam Goudsmit and the Hunt for Hitler's Atom Bomb*, trans. Michiel Horn (Prometheus Books, 2018), 121.
2. Samuel A. Goudsmit, *Alsos* (American Institute of Physics Press, 1996), 112–113.
3. Elisabeth Heisenberg, *Inner Exile: Recollections of a Life with Werner Heisenberg*, trans. S. Cappellari and C. Morris (Birkhäuser, 1984), 108.
4. Thomas Powers, *Heisenberg's War: The Secret History of the German Bomb* (Da Capo Press, 2000), 326.
5. Goudsmit, *Alsos*, 48.

CHAPTER FIFTY-EIGHT. USAAF ATTACK ON TOKYO

1. Malcolm Gladwell, *The Bomber Mafia, a Dream, a Temptation, and the Longest Night of the Second World War* (Little, Brown and Company, 2021), 25.
2. Ibid., 139–140.
3. Ibid., 138, 144.
4. Ibid., 172.
5. John T. Correll, "Daylight Precision Bombing," *Air Force Magazine*, October 1, 2008, 5, 6.
6. Tami Davis Biddle, *Rhetoric and Reality in Air Warfare: The Evolution of British and American Ideas about Strategic Bombing, 1914–1945* (Princeton University Press, 2002), 228, 229.
7. Richard Overy, *The Bombers and the Bombed: Allied Air War Over Europe, 1940–1945* (Penguin, 2013), 139–140.
8. Marshall de Bruhl, *Firestorm: Allied Airpower and the Destruction of Dresden* (Random House, 2006), 226, 283.
9. W. F. Craven and J. L. Cate, *The Army Air Force in World War II*, vol. 5: *The Pacific: Matterhorn to Nagasaki June 1944 to August 1945* (Office of Air Force History, 1983), 617.

10. Lansing Lamont, *Day of Trinity: The Dramatic Story of the Los Alamos Atomic Explosion, and the Men Who, in July 1945, Opened the Nuclear Age* (Antheneum, 1965), 92.
11. Tami Davis Biddle, *Rhetoric and Reality in Air Warfare*, 223.
12. John T. Correll, "The Invasion that Didn't Happen," *Air Force Magazine*, June 1, 2009.
13. Jay Winik, "The 'Frankenstein Weapon,'" *New York Times Book Review*, August 30, 2020.
14. Richard Frank, *Downfall: The End of the Japanese Empire* (Random House, 1999), 342.

CHAPTER FIFTY-NINE. CRITICAL MASS II

1. Stephane Groueff, *Manhattan Project: The Untold Story of the Making of the Atomic Bomb* (iUniverse, 2000), 311.
2. Robert S. Norris, *Racing for the Bomb: General Leslie R. Groves, the Manhattan Project's Indispensable Man* (Steerforth Press, 2002), 371.
3. Erik Durschmid, *Unsung Heroes: The Twentieth Century's Forgotten History-Makers* (F. A. Thorpe Publishing, 2003), 337.
4. Ibid., 340.
5. Lillian Hoddeson, Paul W. Henriksen, Roger A. Meade, and Catherine Westfall, *Critical Assembly: A Technical History of Los Alamos during the Oppenheimer Years, 1943–1945* (Cambridge University Press, 2004), 339.
6. William Lawren, *The General and the Bomb: A Biography of General Leslie R. Groves, Director of the Manhattan Project* (Dodd Mead & Company, 1988), 197.
7. Durschmid, *Unsung Heroes*, 341.
8. Gino Segrè and Bettina Hoerlin, *The Pope of Physics: Enrico Fermi and the Birth of the Atomic Age* (Henry Holt and Company, 2016), 239.
9. Robert Jay Lifton and Gregg Mitchell, *Hiroshima in America: A Half Century of Denial* (Avon Books, 1995), 122.
10. Kai Bird and Martin J. Sherwin, *American Prometheus: The Triumph and Tragedy of J. Robert Oppenheimer* (Vintage Books, 2005), 297.

CHAPTER SIXTY. PREPARING FOR INVASION

1. Charles W. Sweeney with James A. Antonucci and Marion K. Antonucci, *War's End: An Eyewitness Account of America's Last Atomic Mission* (Avon Books, 1997), 236.
2. "Okinawa," *New York Times*, June 5, 1945, 18. "Battle of Okinawa," National WWII

Museum, New Orleans, https://www.nationalww2museum.org/war/topics/battle-of-okinawa

3. "Magic," Far East Summary, War Department, Office of Assistant Chief of Staff G-2, no. 507 (August 9, 1945).
4. "Magic," Far East Summary, War Department, Office of Assistant Chief of Staff G-2, August 4, 1945.
5. J. Samuel Walker, *Prompt and Utter Destruction and the Use of the Atomic Bomb Against Japan*, 3rd ed. (University of North Carolina Press, 2016), 21.

CHAPTER SIXTY-ONE. TRINITY

1. J. Samuel Walker, *Prompt and Utter Destruction and the Use of the Atomic Bomb Against Japan*, 3rd ed. (University of North Carolina Press, 2016), 21, 304.
2. Ralph Nobles, *Trinity: The Day the World Changed, Trinity Atomic Test Site, NM, July 15–16, 1945*, The Atomic Heritage Foundation, https://ahf.nuclearmuseum.org/trinity-day-world-changed-ralph-nobles/.
3. Lansing Lamont, *Day of Trinity: The Dramatic Story of the Los Alamos Atomic Explosion, and the Men Who, In July 1945, Opened the Nuclear Age* (Antheneum, 1965), 1.
4. Ibid., 108, 121–122, 157.
5. Richard Rhodes, *The Making of the Atomic Bomb* (Simon & Schuster, 1986), 659.
6. William Lawren, *The General and the Bomb: A Biography of General Leslie R. Groves, Director of the Manhattan Project* (Dodd Mead & Company, 1988), 208.
7. Lamont, *Day of Trinity*, 177.
8. D. M. Giangreco (author, *Hell to Pay: Operation DOWNFALL and the Invasion of Japan 1945–1947*), interview with NPR's Scott Simon, *Weekend Edition*, January 16, 2010.
9. Lamont, *Day of Trinity*, 180.
10. Kai Bird and Martin J. Sherwin, *American Prometheus: The Triumph and Tragedy of J. Robert Oppenheimer* (Vintage Books, 2005), 264.
11. Bruce Cameron Reed, *The History and Science of the Manhattan Project*, 2nd ed. (Springer-Verlag, 2019), 327.
12. Rhodes, *The Making of the Atomic Bomb*, 661.
13. Lawren, *The General and the Bomb*, 209.
14. Rhodes, *The Making of the Atomic Bomb*, 656.
15. Bird and Sherwin, *American Prometheus*, 21–2, 306.
16. Stephane Groueff, *Manhattan Project: The Untold Story of the Making of the Atomic Bomb*

(iUniverse, 2000), 352.

17. Lamont, *Day of Trinity*, 210.
18. Lamont, *Day of Trinity*, 164.
19. Groueff, *Manhattan Project*, 352.
20. Richard Miller, *Under the Cloud: The Decades of Nuclear Testing* (Two-Sixty Press, 1991), 34.
21. Rhodes, *The Making of the Atomic Bomb*, 666.
22. Lawren, *The General and the Bomb*, 213.
23. Leslie Groves, "Three Years on a Tightrope," in *The Manhattan Project: The Birth of the Atomic Bomb in the Words of Its Creators, Eyewitnesses, and Historians*, ed. Cynthia Kelly (Black Dog & Leventhal Publishers, 2007), 297.
24. Lawren, *The General and the Bomb*, 214.
25. "General Groves Declassified," *Memorandum for the Secretary of War, Subject: The Test*, July 18, 1945.
26. Alex Wellerstein, "The First Light of Trinity," *The New Yorker*, July 16, 2015.
27. Lamont, *Day of Trinity*, 240.
28. "Ammunition Dump in New Mexico Explodes," United Press Archives, July 17, 1945.
29. Lamont, *Day of Trinity*, 237.
30. Lawren, *The General and the Bomb*, 216.
31. William L. Laurence, "Drama of the Atomic Bomb Found Climax in July 16 Test," *New York Times*, September 26, 1945.
32. Lamont, *Day of Trinity*, 244.
33. Ibid., 244, 251.
34. Ibid., 246.
35. Lillian Hoddeson, Paul W. Henriksen, Roger A. Meade, and Catherine Westfall, *Critical Assembly: A Technical History of Los Alamos during the Oppenheimer Years, 1943–1945* (Cambridge University Press, 2004), 373.
36. Lamont, *Day of Trinity*, 248.
37. Ibid., 251.
38. Kathleen M. Tucker and Robert Alvarez, "Trinity: 'The Most Significant Hazard of the Entire Manhattan Project,'" *Bulletin of the Atomic Scientists*, July 15, 2019.
39. Lamont, *Day of Trinity*, 127.
40. Ibid., 187, 193.

CHAPTER SIXTY-TWO. FARM HALL

1. Jeremy Bernstein, *Hitler's Uranium Club: The Secret Recordings at Farm Hall*, 2nd ed. (Copernicus Books, 2001), 319.
2. Elisabeth Heisenberg, *Inner Exile: Recollections of a Life with Werner Heisenberg*, trans. S. Cappellari and C. Morris (Birkhäuser, 1984), 113.
3. Victor Weisskopf, *The Joy of Insight: Passions of a Physicist* (Plunkett Lake Press, 2013).
4. Sam Kean, *The Bastard Brigade: The True Story of the Renegade Scientists and Spies Who Sabotaged the Nazi Atomic Bomb* (Little, Brown and Company, 2019), 405.
5. Bernstein, *Hitler's Uranium Club*, xviii.
6. Ibid., 102, 98.
7. David C. Cassidy, *Uncertainty: The Life and Science of Werner Heisenberg* (W. H. Freeman, 1992), 483.
8. Bernstein, *Hitler's Uranium Club*, 96, 98.
9. Cassidy, *Uncertainty*, 525.
10. Samuel A. Goudsmit, *Alsos* (American Institute of Physics Press, 1996), 125.
11. David Irving, *The Virus House: Germany's Atomic Research and Allied Counter-Measures* (William Kimber, 1967), 275.

CHAPTER SIXTY-THREE. HIROSHIMA

1. John Hersey, *Hiroshima* (Vintage Books, 1973), 2–4.
2. Ibid., 5.
3. Charles W. Sweeney with James A. Antonucci and Marion K. Antonucci, *War's End: An Eyewitness Account of America's Last Atomic Mission* (Avon Books 1997), 168.
4. Jim Baggott, *The First War of Physics: The Secret History of the Atom Bomb 1939–1949* (Pegasus Books, 2010), 334.
5. David Bodanis, *E=MC²: A Biography of the World's Most Famous Equation* (Berkeley Books, 2001), 166–169.
6. The Avalon Project, "The Attacks," in *The Atomic Bombings of Hiroshima and Nagasaki* (Yale Law School, Lillian Goldman Law Library, 2008).
7. David McCullough, *Truman* (Simon & Schuster, 1992), 457.
8. William Lawren, *The General and the Bomb: A Biography of General Leslie R. Groves, Director of the Manhattan Project* (Dodd Mead & Company, 1988), 242.
9. Ibid., 244.

10. Ibid., 244–247.
11. Robert S. Norris, *Racing for the Bomb: General Leslie R. Groves, the Manhattan Project's Indispensable Man* (Steerforth Press, 2002), 357–358.
12. Kai Bird and Martin J. Sherwin, *American Prometheus: The Triumph and Tragedy of J. Robert Oppenheimer* (Vintage Books, 2005), 315.
13. Ibid., 316, 319.
14. Samuel A. Goudsmit, *Alsos* (American Institute of Physics Press, 1996), 131.
15. Daniel Lang, "A Farewell to String and Sealing Wax," in *From Hiroshima to the Moon: Chronicles of Life in the Atomic Age* (Simon & Schuster, 1959), 230.

CHAPTER SIXTY-FOUR. FARM HALL GETS THE NEWS

1. Klaus Hoffmann, *Otto Hahn: Achievement and Responsibility* (Springer-Verlag, 1993), 3.
2. David C. Cassidy, *Beyond Uncertainty: Heisenberg, Quantum Physics, and the Bomb* (Bellevue Literary Press, 2009), 371.
3. Thomas Powers, *Heisenberg's War: The Secret History of the German Bomb* (Da Capo Press, 2000), 437.
4. Leslie R. Groves, *Now It Can Be Told: The Story of the Manhattan Project* (Da Capo Press, 1983), 338.
5. Jean Medawar and David Pyke, *Hitler's Gift: Scientists Who Fled Nazi Germany* (Richard Cohen Books, 2000), 177.
6. Jeremy Bernstein, *Hitler's Uranium Club: The Secret Recordings at Farm Hall*, 2nd ed. (Copernicus Books, 2001), 116.
7. Ibid., 133.
8. Ibid., 116, 125.
9. Samuel A. Goudsmit, *Alsos* (American Institute of Physics Press, 1996), 134–135.
10. Bernstein, *Hitler's Uranium Club*, 122.
11. Cassidy, *Beyond Uncertainty*, 376–378.
12. "Nobel Prizes, Award to a German Scientist," *Guardian*, November 16, 1945.

CHAPTER SIXTY-FIVE. VICTORY

1. John Coster-Mullen, *Atom Bombs: The Top Secret Inside Story of Little Boy and Fat Man* (self-pub., 2009), 41.

2. Richard Rhodes, *The Making of the Atomic Bomb* (Simon & Schuster, 1986), 392.
3. Alex Wellerstein, "How Many People Died at Hiroshima and Nagasaki?," Restricted Data: The Nuclear History Blog, August 4, 2020, https://blog.nuclearsecrecy.com/2020/08/04/how-many-people-died-at-hiroshima-and-nagasaki/.
4. Robert S. Norris, *Racing for the Bomb: General Leslie R. Groves, the Manhattan Project's Indispensable Man* (Steerforth Press, 2002), 416.
5. W. F. Craven and J. L. Cate, *The Army Air Force in World War II*, vol. 5: *The Pacific: Matterhorn to Nagasaki June 1944 to August 1945* (Office of Air Force History, 1983), 617.

CHAPTER SIXTY-SIX. RADIATION POISONING REPORTED

1. William Lawren, *The General and the Bomb: A Biography of General Leslie R. Groves, Director of the Manhattan Project* (Dodd Mead & Company, 1988), 251.
2. "Tokyo Puts Toll of Atomic Bomb At 190,000 Killed and Wounded," *New York Times*, August 23, 1945.
3. Sean Malloy, *"A Very Pleasant Way to Die": Radiation Effects and the Decision to Use the Atomic Bomb Against Japan* (Diplomatic History, 2012), 516.
4. Richard Tanter, "Voice and Silence in the First Nuclear War: Wilfred Burchett and Hiroshima," *The Asia-Pacific Journal* 3, no. 8 (August 3, 2005): 2.
5. Ben Kiernan, ed., *Burchett, Reporting the Other Side of the World 1939–1983* (Quartet Books, 1986), 18, 25.
6. *Daily Express* (London), September 6, 1945, qtd. In Tanter, "Voice and Silence," 7, 4.
7. Kiernan, *Burchett, Reporting*, 18.
8. Malloy, *"A Very Pleasant Way to Die,"* 538.
9. Ibid., 518.
10. Ibid., 541
11. William Laurence, "U.S. Atom Bomb Site Belies," *New York Times*, September 12, 1945.
12. Malloy, *"A Very Pleasant Way to Die,"* 543.
13. Laurence, "U.S. Atom Bomb Site Belies."
14. Manhattan District History, "Chapter 6," in *Investigation of the After Effects of the Bombing in Japan: Book I, General*, vol. 4: *Auxiliary Activities*, 6.29, https://www.osti.gov/opennet/manhattan_district.
15. George E. Jones, "Survey Rules Out Nagasaki Dangers," *New York Times*, October 7, 1945.
16. Manhattan District History, "Chapter 6," in *Investigation of the After Effects of the Bombing in Japan: Book 1, General*, vol. 4: *Auxiliary Activities*, 6.24.

17. Ibid., 530.
18. Alex Wellerstein, "Who Knew about Radiation Sickness and When?," Restricted Data, a Nuclear History Blog, October 18, 2012, https://blog.nuclearsecrecy.com/2012/10/18/who-knew-about-radiation-sickness-and-when.
19. Memorandum from J. R. Oppenheimer to Brigadier General Farrell, May 11, 1945, National Security Archive, https://nsarchive.gwu.edu/document/28511-document-10-memorandum-j-r-oppenheimer-brigadier-general-farrell-may-11-1945.
20. Major General L. R. Groves to Chief of Staff [General George C. Marshall], July 30, 1945, National Security Archive, https://nsarchive.gwu.edu/document/28689-document-11-major-general-lr-groves-chief-staff-general-george-c-marshall-30-july.
21. Malloy, *"A Very Pleasant Way to Die,"* 515–545.
22. Wellerstein, "Who Knew about Radiation Sickness."
23. "Gamma Ray Cause of 20% Of Deaths," *New York Times*, June 30, 1946.

CHAPTER SIXTY-SEVEN. TRINITY FALLOUT

1. Richard Miller, *Under the Cloud: The Decades of Nuclear Testing* (Two-Sixty Press, 1991), 59.
2. Kathleen M. Tucker and Robert Alvarez, "Trinity: 'The Most Significant Hazard of the Entire Manhattan Project,'" *Bulletin of the Atomic Scientists*, 2019, 7–8.
3. Joshua Wheeler, "It's Been 75 Years, and America Still Won't Admit a Nuclear Disaster," *New York Times*, July 15, 2020.
4. "Bomb Story," *New York Times Book Review*, August 30, 2020.
5. Miller, *Under the Cloud*, 59.
6. Tucker and Alvarez, "Trinity."
7. "Study: Cancer Cases Likely in Those Exposed to New Mexico Atomic Test," Associated Press, September 1, 2020.
8. According to a *New York Times* July 20, 2023, article, a recent study determined that "Trinity nuclear test's fallout reached 46 states, Canada and Mexico."
9. "Interview with Karen Dorn Steele," *Voices of the Manhattan Project*, The Atomic Heritage Foundation, https://ahf.nuclearmuseum.org/voices/people/karen-dorn-steele.

EPILOGUE

1. Samuel A. Goudsmit, *Alsos* (American Institute of Physics Press, 1996), 243.
2. Ibid., xiv, xxii.
3. David C. Cassidy, *Beyond Uncertainty: Heisenberg, Quantum Physics, and the Bomb* (Bellevue Literary Press, 2009), 381.
4. David C. Cassidy, *Uncertainty: The Life and Science of Werner Heisenberg* (W. H. Freeman, 1992), 528.
5. Mark Walker, *Nazi Science: Myth, Truth, and the German Atomic Bomb* (Perseus Publishing, 1995), 259.
6. J. Samuel Walker, ed., *Nuclear Energy and the Legacy of Harry S. Truman* (Truman State University Press, 2016), 54.
7. Gino Segrè and Bettina Hoerlin, *The Pope of Physics: Enrico Fermi and the Birth of the Atomic Age* (Henry Holt and Company, 2016), 269.
8. Laura Fermi, *Atoms in the Family: My Life with Enrico Fermi* (University of Chicago Press, 1954), 244.
9. Robert S. Norris, *Racing for the Bomb: General Leslie R. Groves, the Manhattan Project's Indispensable Man* (Steerforth Press, 2002), 461, 465.
10. Ibid., 502–503.
11. Ibid., 503.

BIBLIOGRAPHY

Aczel, Amir D. *Uranium Wars: The Scientific Rivalry That Created the Nuclear Age*. New York: Palgrave Macmillan, 2009.

Asimov, Isaac. *Atom: Journey across the Subatomic Cosmos.* New York: Truman Talley Books/Plume, 1992.

Baggott, Jim. *The First War of Physics: The Secret History of the Atom Bomb 1939–1949*. New York: Pegasus Books, 2010.

Bascomb, Neal. *The Winter Fortress: The Epic Mission to Sabotage Hitler's Atomic Bomb*. Houghton Mifflin Harcourt, 2016.

Bernstein, Jeremy. *Hitler's Uranium Club: The Secret Recordings at Farm Hall.* 2nd ed. New York: Copernicus Books, 2001.

Bird, Kai, and Martin J. Sherwin. *American Prometheus: The Triumph and Tragedy of J. Robert Oppenheimer*. New York: Vintage Books, 2006.

Blume, Lesley M. M. *Fallout: The Hiroshima Cover-up and the Reporter Who Revealed It to the World.* New York: Simon & Schuster, 2020.

Bodanis, David. *E=MC²: A Biography of the World's Most Famous Equation*. New York: Berkeley Books, 2001.

Brode, Bernice. *Tales of Los Alamos: Life on the Mesa 1943–1945*. Los Alamos: Los Alamos Historical Society, 1997.

Cassidy, David C. *Beyond Uncertainty: Heisenberg, Quantum Physics, and the Bomb*. New York: Bellevue Literary Press, 2010.

———. *J. Robert Oppenheimer and the American Century*. Pi Press, 2005.

———. *Uncertainty: The Life and Science of Werner Heisenberg*. New York: W. H. Freeman, 1992.

Churchill, Winston S. *The Second World War: The Gathering Storm*. Boston: Houghton Mifflin, 1976.

Clark, Claudia. *Radium Girls: Women and Industrial Health Reform, 1910–1935*. Chapel Hill: University of North Carolina Press, 1997.

Conant, Jennet. *Man of the Hour: James B. Conant, Warrior Scientist*. New York: Simon & Schuster, 2017.

———. *109 East Palace: Robert Oppenheimer and the Secret City of Los Alamos*. New York: Simon & Schuster, 2005.

———. *Tuxedo Park: A Wall Street Tycoon and the Secret Palace of Science That Changed the Course of World War II*. New York: Simon & Schuster, 2013.

Cornwell, John. *Hitler's Scientists: Science, War and the Devil's Pact*. New York: Penguin Books, 2003.

Coster-Mullen, John. *Atom Bombs: The Top Secret Inside Story of Little Boy and Fat Man*. Self-published, 2009.

Craven, W. F., and J. L. Cate, eds. *The Army Air Force in World War II*. Volume 5, *The Pacific: Matterhorn to Nagasaki June 1944 to August 1945*. Washington, D.C.: Office of Air Force History, 1983.

Daniels, Roger. *Franklin D. Roosevelt: The War Years, 1939–1945*. University of Illinois Press, 2016.

Davis Biddle, Tami. *Rhetoric and Reality in Air Warfare: The Evolution of British and American Ideas about Strategic Bombing, 1914–1945*. Princeton: Princeton University Press, 2002.

De Bruhl, Marshall. *Firestorm: Allied Airpower and the Destruction of Dresden*. New York: Random House, 2006.

Durschmid, Erik. *Unsung Heroes: The Twentieth Century's Forgotten History-Makers*. F. A. Thorpe Publishing, 2003.

Fermi, Enrico, and James W. Cronin. *Fermi Remembered*. Chicago: University of Chicago Press, 1961.

Fermi, Laura. *Atoms in the Family: My Life with Enrico Fermi*. Chicago: University of Chicago Press, 1961.

Fine, Lenore, and Jesse A. Remington. "Atomic Mission," in *The Corps of Engineers: Construction in the United States*, 650–701. Washington, D.C.: Center of Military History, U.S. Army, 1989.

Frank, Richard *Downfall: The End of the Japanese Empire*. Random House, 1999.

Frisch, Otto. *What Little I Remember*. Plunkett Lake Press, 2017.

Gannon, Michael. *Operation Drumbeat: The Dramatic True Story of Germany's First U-Boat Attacks Along the American Coast in World War II*. Harper & Row Publishers, 1990.

Georg, Friedrich. *Hitler's Miracle Weapons*. Volume 1, *Secret Nuclear Weapons of the Third Reich and Their Carrier Systems: The Luftwaffe and Kriegsmarine*. West Midlands, England:

Helion & Company, 2003.

———. *Hitler's Miracle Weapons.* Volume 2, *The History of the Rockets and Flying Craft of the Third Reich, from the V-1 to the A-9, Unconventional Short- and Medium-Range Weapons.* Translated by C. F. Colton MA. West Midlands, England: Helion & Company, 2005.

Gladwell, Malcolm. *The Bomber Mafia: A Dream, a Temptation, and the Longest Night of the Second World War.* New York: Little, Brown and Company, 2021.

Goudsmit, Samuel A. *Alsos.* Woodbury: American Institute of Physics Press, 1996.

Groueff, Stephane. *Manhattan Project: The Untold Story of the Making of the Atomic Bomb.* iUniverse, 2000.

Groves, Leslie R. *Now It Can Be Told: The Story of the Manhattan Project.* New York: Da Capo Press, 1983.

Hachiya, Michihiko. *Hiroshima Diary: The Journal of a Japanese Physician.* Translated and edited by Warner Wells. Chapel Hill: University of North Carolina Press, 1955.

Hallion, Richard. "The Accuracy of Air-Delivered Weapons." Pathfinder, Air Power Development Centre Bulletin, October 2019.

Heisenberg, Elisabeth. *Inner Exile: Recollections of a Life with Werner Heisenberg.* Translated by S. Cappellari and C. Morris. Boston: Birkhäuser, 1984.

Heisenberg, Werner. *Physics and Beyond: Encounters and Conversations.* Translated by Arnold J. Pomerans. New York: Harper & Row, 1972.

Herken, Gregg. *Brotherhood of the Bomb: The Tangled Lives and Loyalties of Robert Oppenheimer, Ernest Lawrence, and Edward Teller.* New York: Henry Holt and Company, 2002.

Hersey, John. *Hiroshima.* Vintage Books, 1989.

Hewlett, Richard G., and Oscar E. Anderson Jr. *A History of the United States Atomic Energy Commission.* Volume 1, *The New World, 1939/1946.* University Park:: Pennsylvania State University Press, 1962.

Hoddeson, Lillian, Paul W. Henriksen, Roger A. Meade, and Catherine Westfall. *Critical Assembly: A Technical History of Los Alamos during the Oppenheimer Years 1943–1945.* Cambridge: Cambridge University Press, 2004.

Hoffmann, Klaus. *Otto Hahn: Achievement and Responsibility.* Berlin: Springer-Verlag, 1993.

Irving, David. *The Virus House: Germany's Atomic Research and Allied Counter-Measures.* William Kimber, 1967.

Isaacson, Walter. *Einstein: His Life and Universe.* Simon & Schuster, 2007.

Johnson, Brian. *The Secret War.* Yorkshire: Pen & Sword Military Classics, 2004.

Jones, R. V. *Most Secret War.* Wordsworth, 1998.

Jones, Vincent C. *Manhattan: The Army and the Atomic Bomb.* Washington, D.C.: Center of

Military History, U.S. Army, 1985.

Joseloff, Michael. *Chasing Heisenberg: The Race for the Atom Bomb*. Amazon Publishing, 2018.

Jungk, Robert. *Brighter Than a Thousand Suns: A Personal History of the Atomic Scientists.* Translated by James Cleugh. New York: Harcourt, Brace, 1958.

Kean, Sam. *The Bastard Brigade: The True Story of the Renegade Scientists and Spies Who Sabotaged the Nazi Atomic Bomb*. New York: Little, Brown and Company, 2019.

Kelly, Cynthia C., ed. *The Manhattan Project: The Birth of the Atomic Bomb in the Words of Its Creators, Eyewitnesses, and Historians*. New York: Black Dog & Leventhal Publishers, 2007.

Kiernan, Ben. *Burchett, Reporting the Other Side of the World, 1939–1983*. London: Quartet Books, 1986.

Kramish, Arnold. *The Griffin: The Greatest Untold Espionage Story of World War II*. Boston: Houghton Mifflin, 1986.

Lamont, Lansing. *Day of Trinity: The Dramatic Story of the Los Alamos Atomic Explosion, and the Men Who, in July 1945, Opened the Nuclear Age*. New York: Antheneum, 1965.

Lang, Daniel. "A Farewell to String and Sealing Wax." In *From Hiroshima to the Moon: Chronicles of Life in the Atomic Age*. New York: Simon & Schuster, 1959.

Lawren, William. *The General and the Bomb: A Biography of General Leslie R. Groves, Director of the Manhattan Project*. New York: Dodd Mead & Company, 1988.

Lewis, Damien. *Hunting the Nazi Bomb: The Special Forces Mission to Sabotage Hitler's Deadliest Weapon*. London: Quercus Editions Ltd, 2017.

Lifton, Robert Jay, and Gregg Mitchell. *Hiroshima in America: A Half Century of Denial*. New York: Avon Books, 1996.

Loring, William S. *Birthplace of the Atomic Bomb: A Complete History of the Trinity Test Site*. Jefferson, NC: Mcfarland & Company, 2019.

MacDonald, Laura M. *Curse of the Narrows: The Halifax Disaster of 1917*. Walker Publishing Company, 2005.

Malloy, Sean. *"A Very Pleasant Way to Die": Radiation Effects and the Decision to Use the Atomic Bomb Against Japan*. Diplomatic History, 2012.

Manchester, William. *The Glory and the Dream: A Narrative History of America, 1932–1972*. Boston: Little, Brown, 1974.

McCullough, David. *Truman*. Simon & Schuster, 1992.

McPhee, John. *The Curve of Binding Energy: A Journey into the Awesome and Alarming World of Theodore B. Taylor*. New York: Farrar, Straus and Giroux, 1977.

Medawar, Jean, and David Pyke. *Hitler's Gift: Scientists Who Fled Nazi Germany*. London: Richard Cohen Books, 2000.

Miller, Richard L. *Under the Cloud: The Decades of Nuclear Testing.* The Woodlands: Two-Sixty

Press, 1991.

Murphy, Mark. *The Exploits of Agent 110: Allen Dulles in Wartime.* CIA Studies in Intelligence, 2013.

Nagorski, Andrew *Hinterland: American Eyewitness to the Nazi Rise to Power*. Simon & Schuster, 2012.

Nobles, Ralph. *Trinity: The Day the World Changed, Trinity Atomic Test Site, NM, July 15–16, 1945*. The Atomic Heritage Foundation, https://ahf.nuclearmuseum.org/trinity-day-world-changed-ralph-nobles/.

Norris, Robert S. *Racing for the Bomb: General Leslie R. Groves, the Manhattan Project's Indispensable Man*. South Royalton: Steerforth Press, 2002.

Overy, Richard. *The Bombers and the Bombed: Allied Air War Over Europe, 1940–1945*. New York: Penguin, 2013.

Pash, Boris T. *The Alsos Mission: The Incredible Story of the Espionage Commandos Who Stole Hitler's Most Lethal Weapon* . New York: Award Books, 1970.

Powers, Thomas. *Heisenberg's War: The Secret History of the German Bomb*. Lebanon, IN: Da Capo Press, 2000.

Price, Alfred. *Instruments of Darkness: The History of Electronic Warfare*. New York: Charles Scribner's Sons, 1978.

Reed, Bruce Cameron. *The History and Science of the Manhattan Project.* 2nd ed. Springer-Verlag, 2019.

Rhodes, Richard. *The Making of the Atomic Bomb*. New York: Simon & Schuster, 1986.

Richelson, Jeffrey T. *Spying on the Bomb: American Nuclear Intelligence from Nazi Germany to Iran and North Korea*. New York: W. W. Norton & Company, 2006.

Rose, Paul Lawrence. *Heisenberg and the Nazi Atomic Bomb Project: A Study in German Culture.* Berkeley: University of California Press, 1998.

Rotter, Andrew J. *Hiroshima: The World's Bomb*. Oxford: Oxford University Press, 2008.

Sanger, S. L., and Craig Wollner. *Working on the Bomb: An Oral History of WWII Hanford.* Portland: Continuing Education Press, Portland State University, 1995.

Schwartz, David N. *The Last Man Who Knew Everything: The Life and Times of Enrico Fermi, Father of the Nuclear Age*. New York: Basic Books, 2017.

Seaborg, Glenn T., and Eric Seaborg. *Adventures in the Atomic Age: From Watts to Washington.* New York: Farrar, Straus and Giroux, 2001.

Segrè, Gino, and Bettina Hoerlin. *The Pope of Physics: Enrico Fermi and the Birth of the Atomic Age*. New York: Henry Holt and Company, 2016.

Serber, Robert. *The Los Alamos Primer: The First Lectures on How to Build an Atomic Bomb.* Berkeley: University of California Press, 1992.

Shirer, William L. *The Rise and Fall of the Third Reich: A History of Nazi Germany*. New York: Simon & Schuster, 2011.

Smyth, Henry DeWolf. "Atomic Energy for Military Purposes." *The Smyth Report*, July 1945, https://www.atomicarchive.com/resources/documents/smyth-report/index.html.

Speer, Albert. *Inside the Third Reich: Memoirs*. Translated by Richard and Clara Winston. New York: Simon & Schuster, 1970.

Stansky, Peter. *The First Day of the Blitz: September 7, 1940*. Yale University Press, 2007.

Sweeney, Charles W., with James A. Antonucci and Marion K. Antonucci. *War's End: An Eyewitness Account of America's Last Atomic Mission*. New York: Avon Books, 1997.

Szasz, Ferenc Morton. *The Day the Sun Rose Twice: The Story of the Trinity Site Nuclear Explosion, July 16, 1945*. Albuquerque: University of New Mexico Press, 1984.

Tanter, Richard. "Voice and Silence in the First Nuclear War: Wilfred Burchett and Hiroshima." *The Asia-Pacific Journal* 3, no. 8 (August 3, 2005), https://apjjf.org/richard-tanter/2066/article.

van Calmthout, Martijn. *Sam Goudsmit and the Hunt for Hitler's Atom Bomb*. Translated by Michiel Horn. New York: Prometheus Books, 2018.

Von Schirach, Richard. *The Night of the Physicists: Operation Epsilon: Heisenberg, Hahn, Weizsäcker and the German Bomb*. Translated by Simon Pare. London: Haus Publishing, 2014.

Walker, Mark. *Nazi Science: Myth, Truth, and the German Atomic Bomb*. Cambridge, MA: Perseus Publishing, 1995.

Walker, J. Samuel, ed. *Nuclear Energy and the Legacy of Harry S. Truman*. Kirksville, MO: Truman State University Press, 2016.

———. *Prompt and Utter Destruction and the Use of the Atomic Bomb against Japan*. 3rd ed. Chapel Hill: University of North Carolina Press, 2016.

Weisskopf, Victor. *The Joy of Insight: Passions of a Physicist*. Plunkett Lake Press, 2013.

Wells, H. G. *The World Set Free*. Cabin John, MD: Wildside Press LLC.

INDEX

A

C

D

E

F

G

H

I

J

K

L

M

N

O

P

Q

R

U

V

W

X